Traditional Phytotherapy

The Editor

Dr. Jitu Buragohain obtained M.Sc. (Botany) from Dibrugarh University, Assam and Ph. D. in Plant Biotechnology from Tezpur University, Assam. Currently, he is working as the Principal of Namrup College, Dist. Dibrugarh, Assam. He has published a number of research papers in National and International journals of repute including Elsevier and Springer and also authored two books.

Traditional Phytotherapy

Editor

Jitu Buragohain

2013

Daya Publishing House®

A Division of

Astral International Pvt. Ltd.

New Delhi – 110 002

Published by   :   **Daya Publishing House®**
                            A Division of
                            Astral International Pvt. Ltd.
                            ISO 9001:2008 Certified Company
                            4760-61/23, Ansari Road, Darya Ganj
                            New Delhi-110 002
                            Ph. 011-43549197, 23278134
                            E-mail: info@astralint.com
                            Website: www.astralint.com

Laser Typesetting   :   **Classic Computer Services**
                            Delhi - 110 035

Printed at   :   **Salasar Imaging Systems**
                            Delhi - 110 035

PRINTED IN INDIA

Preface

The North Eastern region of India is recognized as one of the mega biodiversity centers of the world. The area lies between 22°-30° N latitude and 89°-97° E longitude, and spreading over 2,62,379 sq km, represents the transition zone between Indian, Indo-Malayan and Indo-Chinese biogeographic regions and a meeting place of the Himalayan Mountains and Peninsular India. The region possesses more than 2,000 medicinal and aromatic plant species accounting for about 20 per cent of the total plant diversity of India. A large portion of the population of this region still relies mainly on traditional herbal practitioners and local medicinal plants to satisfy their primary health care needs. Research and training activities with regard to traditional medicine of the region has not received due support and attention. As a result, the quantity and quality of safety and efficacy data are far from sufficient to meet the demands for the use of traditional medicine. Safety and efficacy data exist only in respect of much smaller number of plants and their extracts and active ingredients. Reasons for the lack of research data involve not only in policy problems, but also in the research methodology for evaluating traditional medicine. Thus, there is an urgent need for validation and standardization of traditional medical practices of North East India, so that this sector can be accorded its rightful place in the health care system. Patenting and IPR issues are major concern in which the people of North East India need to think very cautiously if medicinal resources are to be commercialized for the welfare of this region. Keeping these views in mind a national seminar was organized with financial assistance from UGC on the theme "*Recent Trends in Traditional Phytotherapy: Safety, Efficacy, Drug Discovery and Priority Issues*" on 2nd and 3rd December, 2010 at Namrup College, Assam, India to promote coordinated efforts for evaluation of traditional medicine in an effective manner. A total of 80 papers had been received by the organizing committee for the seminar and 46 scientific papers had been deliberated by researchers of different places of North East India. This volume is the compilation of 18 selected full papers contributed by eminent scientific and teachers in their respective field. The book is expected to help the students, researchers, NGOs and policy planners.

Many thanks are due to Mr. Sanjay Kumar Seal, Ex-Principal i/c, Namrup College; Prof. Sarada Kanta Sarma, Department of Botany, Gauhati University; Dr. Manabjyoti Bordoloi, Scientist E-II, North East Institute of Science and Technology, Jorhat (CSIR); Dr. Hui Tag, Department of Botany, Rajiv Gandhi University, Arunachal Pradesh; Dr. Bishnu Prasad Sarma, Govt. Ayurvedic College, Guwahati; Dr. Lakhi Ram Saikia, Department of Life Sciences, Dibrugarh University; Dr. Bijay Neog, Department of Life Sciences, Dibrugarh University; Dr. Partha Pratim Baruah, Department of Botany, Gauhati University; Mr. Ghana Buragohain, Ex-Principal i/c, Namrup College; Mr. Bimal Chandra Gogoi, Head, Department of Botany, Namrup College; Mr. Sanjay Dutta, Chairman, Assam Branch Indian Tea Association, Naharkatia Circle; Namrup College Teachers' Unit; Brahmaputra Valley Fertilizer Corporation Ltd, Parbatpur, Namrup and Assam Petrochemicals Ltd, Parbatpur, Namrup for their timely support, advices and suggestions rendered for successfully holding the national seminar at Namrup College. The financial support from Oil India Limited, Duliajan in bringing out this book is gratefully acknowledged. Without the support from OIL, Duliajan it would not have been possible to publish this book in the present form.

Jitu Buragohain

Content

List of Contributors

Acharyya, Padma
Associate Professor, Department of Chemistry, Namrup College, Assam

Baruah, Indira
Assistant Professor, Department of Economics, Namrup College, Assam

Baruah, N.C.
Deputy Director, NPC Division, NEIST, Jorhat, Assam

Bezbaruah, R.L.
North-East Institute of Science and Technology (CSIR), Jorhat, Assam

Bhattacharyya, Nizara
Department of Botany, Sibsagar College, Joysagar

Borah, Rajib Lochan
Assistant Professor, Department of Botany, D.H.S.K. College. Dibrugarh – 786 001, Assam.

Bordoloi, M.J.
North-East Institute of Science and Technology (CSIR), Jorhat, Assam

Bordoloi, Manobjyoti
Natural Product Chemistry Division, North East Institute of Science and Technology, Jorhat – 785 006, Assam

Buragohain, J.
Department of Botany, Namrup College, Namrup, Assam

Chowdhury, D.
Krishi Vigyan Kendra, Lakhimpur

Das, A.K.
Plant Systematic and Pharmacognosy Research Laboratory, Department of Botany, Rajiv Gandhi University, Rono Hills, Doimukh, Itanagar – 791 112, Arunachal Pradesh

Das, Chandan
Associate Professor, Department of Botany, Duliajan College

Das, Manika
Department of Biotechnology, Darrang College, Tezpur – 784 001, Assam

Gogoi, Annajyoti
Associate Professor, Botany Department, Dhemaji College, Dhemaji

Gogoi, Dhrubajyoti
Bioinformatics Infrastructure Facility (DBT-BIF), Department of Botany, Rajiv Gandhi University, Itanagar, Arunachal Predesh

Gogoi, Durga Prasad
Associate Professor, Department of Physics, Namrup College, Assam

Gogoi, Tulika
Lecturer, Botany Department, L.T.K. College, N. Lakhimpur

Gohain (Neog), Jully
Lecturer, Botany Department, L.T.K. College, N. Lakhimpur

Goswami, R.K.
Department of Crop Physiology, BNCA, AAU, Biswanath Chariali

Hazarika, Ridip
Department of Life Sciences, Dibrugarh University, Assam

Kalita, J.C.
Department of Zoology, Gauhati University, Guwahati – 781 014, Assam

Kalita, P.
Plant Systematic and Pharmacognosy Research Laboratory, Department of Botany, Rajiv Gandhi University, Rono Hills, Doimukh, Itanagar – 791 112, Arunachal Pradesh

Kandali, R.
Department of Biochemistry and Agril. Chemistry, BNCA, AAU, Biswanath Chariali

Kangabam, R.D.
Bioinformatics Infrastructure Facility (DBT-BIF), Department of Botany, Rajiv Gandhi University, Itanagar, Arunachal Predesh

Khatun, Firoza
Associate Professor, Department of Botany, Namrup College, Assam

Konwar, B.K.
Department of Molecular Biology and Biotechnology, Tezpur University, Napam, Tezpur Assam

Konwar, Mitali
Department of Physics, Moran College, Moran, Sibsagar – 765 670, Assam

Kotoky, J.
Institute of Advanced Study in Science and Technology, Guwahati – 781 035, Assam

Mudoi, Pronab
Department of Molecular Biology and Biotechnology, Tezpur University, Tezpur, Assam

Neog, Bijay
Department of Life Sciences, Dibrugarh University, Assam

Pradeep, Salam
Bioinformatics Infrastructure Facility (DBT-BIF), Tezpur University, Tezpur, Assam

Prasad, Surya Bali
Cell and Tumor Biology Laboratory, Department of Zoology, North-Eastern Hill University, Shillong – 793 022

Rethy, Parkkal
Department of Forestry, North Eastern Regional Institute of Science and Technology Nirjuli, Arunachal Pradesh

Saikia, Deepjyoti
Associate Professor, Department of Botany, Duliajan College

Saikia, Prabhat
Lecturer, Botany Department, L.T.K. College, N. Lakhimpur

Sharma, K.K.
Institute of Advanced Study in Science and Technology, Guwahati – 781 035, Assam

Singh, Binay
Department of Forestry, North Eastern Regional Institute of Science and Technology Nirjuli, Arunachal Pradesh

Singh, R.K.
Bioinformatics Infrastructure Facility (DBT-BIF), Department of Botany, Rajiv Gandhi University, Itanagar, Arunachal Predesh

Tag, Hui
Plant Systematic and Pharmacognosy Research Laboratory, Department of Botany, Rajiv Gandhi University, Rono Hills, Doimukh, Itanagar – 791 112, Arunachal Pradesh

Verma, Akalesh Kumar
Cell and Tumor Biology Laboratory, Department of Zoology, North-Eastern Hill University, Shillong – 793 022

Chapter 1

Studies on the Antidermatophytic Activity of *Ranunculus sceleratus* Linn. and *Pongamia pinnata* Pierre.

K.K. Sharma, J. Kotoky and J.C. Kalita

ABSTRACT

Dermatophyte fungi are the main agents of skin diseases of man and animal and remain a threat to public health in the tropics. Although a large number of synthetic allopathic drugs are present to treat dermatophytosis, the increasing incidences of fungal resistance towards these synthetic drugs combined with their associated side effects and limited efficacy has forced scientists to search for new antimicrobial substances. Based on the local use of some plants against common diseases and Ethnobotanical knowledge, an attempt has been made to assess the antidermatophytic properties of the organic extracts of two plants Ranunculus sceleratus Linn. and Pongamia pinnata Pierre readily available in Assam, against four species of dermatophytes *viz*. Trichophyton mentagrophytes, Trichophyton rubrum, Microsporum gypseum and Microsporum fulvum using agar cup diffusion technique. The minimum inhibitory concentration of the extracts was determined using two fold serial dilution method. The evaluated results show that these plants possess antifungal properties against the tested dermatophytes.

Keywords: Antidermatophytic, Dermatophytosis, Anti-fungal, Medicinal plants.

Introduction

Dermatophyte fungi are the main agents of skin diseases of man and animal and remain a threat to public health in the tropics. Dermatophytoses or Ringworm are common superficial cutaneous fungal infections caused by filamentous fungi such as *Trichophyton, Microsporum* or *Epidermophyton* species (Kane *et al.,* 1999) which have the capacity to invade keratinous tissues, such as hair, skin or nails, of humans and other animals (Weitzman *et al.,* 1995). Although a large number of synthetic allopathic drugs are present to treat dermatophytosis, the increasing incidences of fungal resistance towards these synthetic drugs combined with their associated side effects like gastrointestinal disturbances, cutaneous reaction, hepatotoxicity and leucopenia in some of the treated patients and limited efficacy has forced scientists to search for new antimicrobial substances (Aguila *et al.,* 1992; Torok *et al.,* 1993; Lopez-Gomez *et al.,* 1994; Gupta *et al.,* 1998).

Medicinal plants are the oldest known health-care products. Natural products have served as a major source of drugs for centuries, and about half the pharmaceuticals in use today are derived from natural products (Clark, 1996). One study has reported that 25 to 50 per cent of current pharmaceuticals are derived from plants. Microbiologists and natural-product chemists are trying to discover more about phytochemicals, which could be developed for treatment of infectious diseases (Cowan, 1999) As opposed to synthetic drugs, antimicrobials of plant origin are not associated with many adverse effects and have an enormous therapeutic potential to heal many infectious diseases (Khan *et al.,* 2010). Based on the local use of some plants against common diseases and ethnobotanical knowledge, an attempt has been made to assess the antidermatophytic properties of two plants *Ranunculus sceleratus* Linn. (Family- Ranunculaceae) and *Pongamia pinnata* Pierre (Family-

Fabaceae) readily available in Assam, against four species of dermatophytes *viz. Trichophyton mentagrophytes, Trichophyton rubrum, Microsporum gypseum* and *Microsporum fulvum* using agar cup diffusion technique.

Materials and Methods

Collection, Identification and Preparation of Plant Materials

The leaves and young shoots of the two plants were collected from Kamrup district, (25.43° and 26.51° North Latitude and between 90.36° and 92.12° East Longitude) of Assam, India. The plant materials were authenticated by a taxonomist from the Gauhati University Assam (India). Herbarium voucher specimen were prepared and deposited in the Life Sciences Division of IASST, Guwahati, Assam, India for future reference.

Collection of Fungal strains

Fungal strains were procured from Institute of Microbial Technology (IMTECH), Chandigarh-160036 (India). The organisms tested were *Trichophyton rubrum* (MTCC 8477), *Trichophyton mentagrophytes* (MTCC 8476), *Microsporum gypseum* (MTCC 8469) and *Microsporum fulvum* (MTCC 8478). The procured samples were sub cultured and maintained in Sabouraud Dextrose Agar (HIMEDIA) slants at 4°C.

Plant Extracts Oreparation

Freshly collected plant materials were washed twice with distilled water to remove the sand and dirt. Plant materials were then dried under shade in a well ventilated room and dried parts of the plants were finely powdered in a mixture grinder. The powdered materials were exhaustively extracted with chloroform and methanol. In brief, about 500 gm of the dried powdered plant materials were dissolved in 1000 ml of solvent for 48 hours for chloroform and methanol extract and repeated for two times. The filtrate is then concentrated under reduced pressure using a rotary evaporator (Buchi R-124) at low temperature (< 40 °C). Finally vacuum desiccators were used to completely remove the solvent. The extracted samples were kept in refrigerator at 4°C for future use (Duraipandiyan *et al.*, 2006).

In vitro Assay

The antifungal activity of the test extracts was determined by employing agar cup diffusion technique (Motiejunaite *et al.*, 2004). A control set was maintained with DMSO. Clotrimazole was used as a reference standard. The plates were incubated at 28 ± 2°C for 24 h to 2 weeks depending on the growth rate of the test pathogens. The experiment was replicated thrice and the average results were recorded. The antifungal activities of the extracts were determined by measuring the diameter of the inhibition zone around the well that was filled with the extract

Determination of Minimum Inhibitory Concentration

Determination of the Minimum Inhibitory Concentration (MIC) was carried out of all the extract that showed inhibitory effect on the test micro-organisms. Two fold serial dilution method was used (Irobi *et al.*, 1993) to determine the MIC.

Results and Discussion

The inhibition zone for the chloroform and methanol extracts of the two plants at 10mg/ml are shown in Figures 1.1–1.4 as obtained from the Agar well diffusion technique. Table 1.1 represents the MIC values of the extract against the tested dermatophytes. The highest inhibition zone of 23 mm was found for the chloroform extract of *R. sceleratus* against *T. mentagrophytes.* The minimum inhibitory concentration of the different extracts against the tested dermatophytes was found ranging between 1.25-10 mg/ml.

Table 1.1: Minimum Inhibitory Concentration for the Different Extracts (mg/ml)

Plants	T. mentagrophytes	T. rubrum	M. gypseum	M. fulvum
R. sceleratus (Chloroform)	1.25-2.5	2.5-5.0	2.5-5.0	2.5-5.0
R. sceleratus (Methanol)	2.5-5.0	2.5-5.0	2.5-5.0	2.5-5.0
P. pinnata (Chloroform)	2.5-5.0	2.5-5.0	2.5-5.0	2.5-5.0
P. pinnata (Methanol)	2.5-5.0	2.5-5.0	5.0-10.0	2.5-5.0

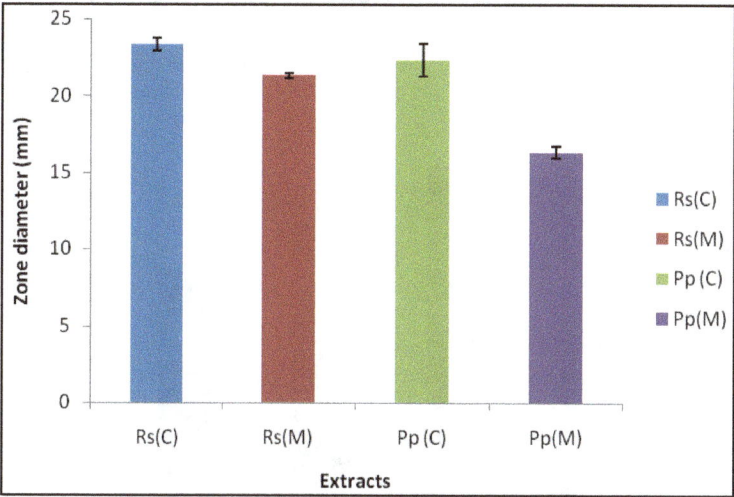

Figure 1.1: Inhibition Zone of Different Extracts Against *T. mentagrophytes*

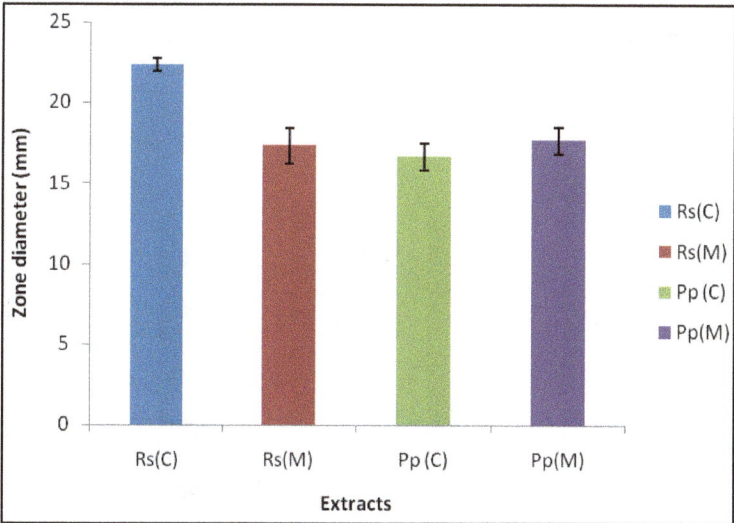

Figure 1.2: Inhibition Zone of Different Extracts Against *T. rubrum*

With the rise in the emergence of various multi drug resistant microorganisms and the scenario worsening through the indiscriminate use of antibiotics, new and/or alternative

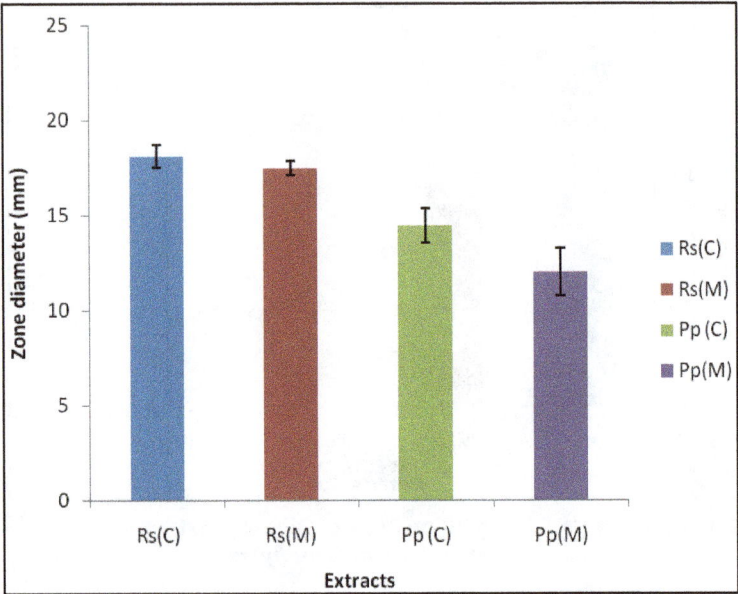

Figure 1.3: Inhibition Zone of Different Extracts Against *M. fulvum*

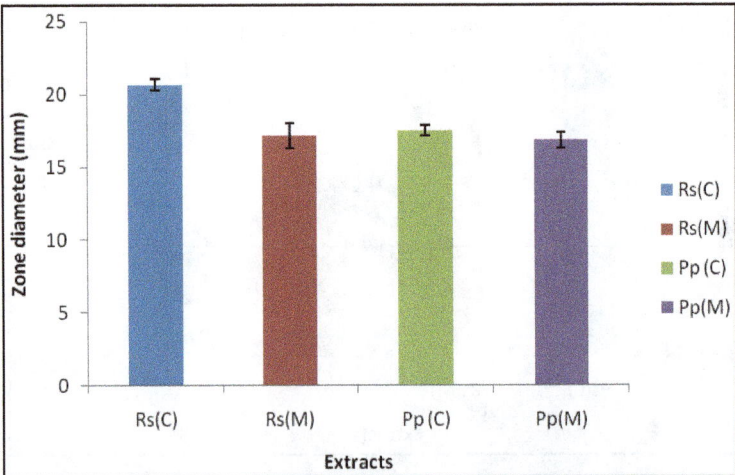

Figure 1.4: Inhibition Zone of Different Extracts Against *M. gypseum*

antimicrobial compounds must be developed to treat common infections. With the changing patterns of susceptibility and the availability of new antimicrobial agents, continuous updating of knowledge concerning treatment of disease caused by such pathogens is required (Khan *et al.*, 2010). In the present study the antifungal activity of the Chloroform and Methanol extracts of the two plants *Ranunculus sceleratus and Pongamia pinnata* against the four dermatophyte strains was evaluated. These findings indicated the presence of active anti-microbial agents in the plant thereby justifying their use for treatment of microbial infections. It is evident from the results that the chloroform extract of *R. sceleratus* showed better activity and is most active of all the extracts. Our results demonstrated that antifungal activities of the leaf extract of *R. sceleratus and P. pinnata* has concentration dependent anti-fungal activity against all tested strains. All the methods used, showed that *Tricophyton mentagrophytes* was the most susceptible strain to the chloroform leaf extracts of *R. sceleratus*

As expected the crude extracts of the chloroform and methanolic leaf extract of *of both the plants* with many components, have a much larger MICs value than the standard drug clotrimazole. However as these plants are highly used by the ethno biologist against a large number of diseases it is expected to have a very low toxicity and one might conclude that the use of these plants would probably produce less side-effects and toxicity compared with conventional chemotherapeutic agents.

Conclusion

The results of the present study support the folkloric usage of the studied plants and suggest that these plant extracts possess compounds with antimicrobial properties that can be further explored for antimicrobial activity. This antifungal study of the plant extracts demonstrated that folk medicine can be an alternative effective way to combat pathogenic microorganisms.

References

Cowan, M.M., 1999. Plant products as antimicrobial agents. *Clin. Microbiol. Rev.* 12: 564–582.

Del Aguila, R., Gei, F.M., Robles, M., Perera-Ramirez, A. and Male, O., 1992. Once-weekly oral doses of fluconazole 150 mg in the treatment of tinea pedis. *Clin. Exp. Dermatol.*, 17: 402–406.

Duraipandiyan, V., Ayyanar, M. and Ignacimuthu, S., 2006. Antimicrobial activity of some ethno-medicinal plants used by

Paliyar tribe from Tamil Nadu, India. *BMC Complementary and Alternative Medicine,* 6: 35.

Gupta, A.K., Del Rosso, J.Q., Lynde, C.W., Brown, G.H. and Shear, N.H., 1998. Hepatitis associated with terbinafine therapy: Three case reports and a review of the literature. *Clin. Exp. Dermatol.,* 23: 64–67.

Irobi, O.N. and Daramola, S.O., 1993. Antifungal activities of crude extracts of *Mictracarpus villosus* (Rubiaceae). *Journal of Ethnopharmacology,* 40: 137–140.

Kane, J. and Summerbel, R.C. Trychophyton, Microsporum, Epidermophyton, and agents of superficial mycoses. In: Murray, Baron PR, Pfaller EJ, Tenover MA.

Khan, R., Zakir, M., Afaq, S.H., Latif, A. and Khan, A.U., 2010. Activity of solvent extracts of *Prosopis spicigera, Zingiber officinale* and *Trachyspermum ammi* against multidrug resistant bacterial and fungal strains. *J. Infect. Dev. Ctries,* 4(5): 292–300.

Lopez-Gomez, S., Del Palacio, A., Van Gutsem, J., Soledad Cuetara, M., Igalesias, L. and Rodriguez-Noriega, A., 1994. Itraconazole versus Griseofulvin in the treatment of tinea capitis: A double blind randomised study in children. *Int. J. Dermatol.,* 33: 743–747.

Motiejunaite, O. and Peciulyte, D., 2004. Fungicidal properties of *Pinus sylvestris* L. for improvement of air quality. *Medicina (Kaunas),* 40(8): 787–793.

Torok, I. and Stehlich, G., 1993. Double blind comparative examination of Ketokonazole 1 per cent cream and Clotrimazole 2 per cent ointment in superficial dermatomycosis. *Ther. Hung.,* 41: 60–63.

Weitzman, I. and Summerbell, R.C., 1995. The dermatophytes. *Clin. Microbiol. Rev.,* 8: 240–259.

Chapter 2

Ligand-Based Virtual Screening on some of the Anticancer Phytochemicals to Develop a Novel Inhibitor of β-catenin against Cancer

Dhrubajyoti Gogoi*, A.K. Verma, R.K. Singh,
M.J. Bordoloi, R.L. Bezbaruah, R.D. Kangabam,
Salam Pradeep and B.K. Konwar

ABSTRACT

Traditionally used plant derived compounds have been a vital source of varied clinically useful anti cancer agents. Recently, numerous bioactive anti cancer compounds have been isolated from plant and animal sources and many of them are currently in clinical trials. Some of the notable published potential anti cancer phytochemicals such as Topotecan, Irinotecan, Etoposide, Paclitaxel (taxol),Vinblastine, Vincristine, Flavopiridol, Rohitkine, Cantharidin etc. has shown good

*   Corresponding author: E-mail: dhruba.bio.du@gmail.com

anticancer activity against different cancer cell lines. The recent study dealt with the in silico investigation of best inhibitor of β-catenin, a well known cancer target. The specific target of Wnt/β-catenin signaling known today is c-myc and cyclin-D1 which is associated with proliferation of cancer cells. In this current work the crystallographic structure of Human Tcf-4/β-Catenin (PDB ID: 1JPW) was retrieved from Protein Data Bank and docked using MVD against anticancer phytochemicals retrieved from Indian Plant Anticancer Compounds Database (InPACdb). Finally five ligands were selected based on best docking score and submitted to the online PharmaGist server to developed pharmacophore model and pivot molecule. The Physiochemical Properties, Bioactivity and Toxicity were predicted for each compound which showed best docking. This work will be helpful in development of potent inhibitor of β-catenin and also it will give a clue to synthesize this inhibitor in chemical laboratory for further evaluation of anticancer activity against different cancer cell lines.

Keywords: *β-catenin, Docking, Anticancer, Phytochemicals, Toxicity, Cancer cells.*

Introduction

Cancer is an abnormal growth of cells caused by multiple changes in gene expression leading to dysregulated balance of cell proliferation and cell death and ultimately evolving into a population of cells that can invade tissues and metastasize to distant sites, causing significant morbidity and, if untreated, death of the host. Most of the proteins encoded by tumor suppression genes act as negative regulators of cell proliferation (Prasad, 2005), which may be as transcription factors (p53 and WTI), cell cycle regulators (RB and p16), components regulating signalling pathways (NFI), regulating RNA polymerase II elongation (VHL). Thus, their elimination contributes and promotes uncontrolled cell growth (Haber and Harlow, 1997).

Nature has provided many things for humankind over the years, and the use of medicinal plants and animals for health reasons started thousands of years ago and is still part of medical practice in many developing countries. Chinese herb guides document the use of herbaceous plants as far back in time as 2000 BC. (Holt and

Chandra, 2002). Egyptians have been found to have documented uses of various herbs in 1500 BC (Cragg and Newman, 2001). World Health Organization estimates that approximately 80 percent of the world's population relies primarily on traditional medicines as sources for their primary health care (Farnsworth *et al., 1985*). Since 1961, over nine plant-derived compounds have been approved for use as anticancer drugs in the US. Some of these compounds are: vinblastine, vincristine, Navelbine, etoposide, teniposide, Taxol (paclitaxel), Taxotere (docetaxel), topotecan and irinotecan (Kuo-Hsiung, 1999). Podophyllin extracted from *Podophyllum peltatum* has also been found to inhibit mitosis *in vitro* and its derivatives are found to be capable of arresting cells in either late S phase or early G2 phase, without inhibiting microtubule assembly. Taxol and docetaxel derived from *Taxus brevifolia* and *Taxus baccata* respectively are also active in preclinical animal screening systems for anticancer drugs (Sllchenmyer and Von Hoff, 1991). Turmeric extract and curcumin isolated from it is effective in reducing animal tumors, indicating its potential for use in cancer treatment (Krishnaswamy *et al., 1998*).

In Asia, North-East India is one of the biodiversity hotspots. The region is endowed with varied flora to its diversified topography and climatic conditions marked by high rainfall, moderate temperature and high humidity. Different tribes living in this area mostly rely on traditional herbal medicine for their primary healthcare practices. various plants like *Claoxylon hassianum, Celerodendrum wallichii, Mussaenda macrophylla, Blumea lanceolaria, Dillenia pentagyna, Ageratum conyzoides,* etc., are traditionally used by the tribal people of Mizoram (Lalramnghinglova, 1999), and *Taxus baccata, Potentilla fulgens, Panax pseudoginseng,* etc., are used by the tribal people of Meghalaya (Rosangkima and Prasad, 2004).

Our preliminary investigation through literature search showed that Wnt/β-catenin pathway plays a critical role in regulating the growth and maintenance of different cancer cells (Clever, 2006 and Moon *et al.,* 2004). Beta-catenin (or β-catenin) is a protein that in humans is encoded by the *CTNNB1* gene. β-catenin is a subunit of the cadherin protein complex and has been implicated as an integral component in the Wnt signaling pathway (Jansson *et al.,* 2005 and Botrugno *et al.,* 2004). When β-catenin was sequenced it was found to be a member of the armadillo family of proteins. These proteins have multiple copies of the so-called armadillo repeat domain which is specialized for protein-protein binding. When β-catenin is not associated with cadherins and alpha-catenin, it can interact with other proteins such as ICAT and APC. When Wnt is not present, GSK-3 (a kinase) constitutively phosphorylates the β-catenin protein. β-catenin is associated with axin (scaffolding protein) complexed with GSK3 and APC (adenomatosis polyposis coli). The creation of said complex acts to substantially increase the phosphorylation of β-catenin by facilitating the action of GSK3. When β-catenin is phosphorylated it is degraded and thus will not build up in the cell to a significant level. When Wnt binds to frizzled (Fz), its receptor, dishevelled (Dsh) is recruited to the membrane. GSK3 is inhibited by the activation of Dsh by Fz. Because of this, β-catenin is permitted to build up in the cytosol and can be subsequently translocated into the nucleus to perform a variety of functions. It can act in conjunction with TCF and LEF to activate specific target genes like c-myc and Cyclin-D involved in cancer cells growth and proliferation (Nath *et al.,* 2003 and Singh *et al.,* 2006).

Therefore, the recent computer assisted virtual screening method was carried out to develop a potent drug candidate that can inhibit β-catenin. It is suggested that putative pivot molecule from this screening may have promising uses in cancer therapy.

Materials and Methods

Compound Library Retrieval

Compounds (125) were retrieved from Indian Plant Anticancer Compounds Database (InPACdb,), a database of potent anti cancer phytochemicals of Indian Origin. InPACdb's search engine is based on Google's proprietary search technology which utilizes java scripting and custom indexing; making search comparatively faster and user friendly. The database presently comprises a dataset of 144 compounds which has been manually sorted from 990 compounds; 200 from National Cancer institute drug list and 690 from Asian anticancer material databank and other 100 compounds from other miscellaneous sources. The target protein (PDB ID: 1JPW (2.50 Å) - Crystal Structure of a Human Tcf-4/β-Catenin Complex) has been collected from Protein Data Bank.

Tools/Cheminformatics Resources

1. Protein Databank: www.pdb.org
2. InPACdb: www.inpacdb.org
3. OpenBable 2.2.3

Physiochemical Property Calculation

Physiochemical Chemistry or Physical chemistry is the study of macroscopic, atomic, subatomic, and particulate phenomena in chemical systems in terms of physical concepts; often using the principles, practices and concepts of physics like thermodynamics, quantum chemistry, statistical mechanics and dynamics. Physiochemical property such as Lipinski Rule, is one of the most used filters.

Lipinski's rule says that, in general, an orally active drug has no more than one violation of the following criteria:

1. Not more than 5 hydrogen bond donors (nitrogen or oxygen atoms with one or more hydrogen atoms).
2. Not more than 10 hydrogen bond acceptors (nitrogen or oxygen atoms).
3. A molecular weight under 500 Daltons.
4. An octanol-water partition coefficient log P of less than 5.

The partition coefficient is a ratio of concentrations of un-ionized compound between the two solutions. To measure the partition coefficient of ionizable solutes, the pH of the aqueous phase is adjusted such that the predominant form of the compound is un-ionized. The logarithm of the ratio of the concentrations of the un-ionized solute in the solvents is called log P (C.A. Lipinski *et al.*).

Tool: Molsoft Browser 3.6

Bioactivity and Toxicity Prediction using PASS

PASS (Prediction of Activity Spectra of Substances) is software for predicting the biological activity spectra for chemical substances on the basis of their structural formulae, Structure Activity Relationship data and knowledge base. The chemical information must be represented as SD files (format for MDL Information System, Inc), which can be exported from many Chemical Database

Management System. It is developed as a tool for evaluation of general biological potential in a molecule under study.

Energy Minimization

Energy Minimization (energy optimization) methods are common techniques to compute the equilibrium configuration of molecules. The basic idea is that a stable state of a molecular system should correspond to local minima of their potential energy.

When a model is built, the location for each atom may not accurately represent the atom's location in the actual molecule. The model may depict high energy strain at various bonds or conformational strain between atoms. To correct the model one may consider performing an energy minimization. In this aspect we used the MM2.

MM2 (Merck Molecular -2)

MM2 is a family of force field method developed my Merck Research Laboratories. MM2 is most commonly used for energy minimization and calculating properties of organic molecular models.

Tool: ChemBio Office 2010 (Chem3D 12.0)

Protein Setup, Cavity Detection and Molecular Docking

Molecular docking was carried out with Molegro Virtual Docker (MVD) 4.02. MVD is based on differential evolution algorithm; the solution of the algorithm takes into account the sum of the intermolecular interaction energy between the Ligand and the protein, and the intramolecular interaction energy of the Ligand. The docking energy scoring function is based on a modified piecewise linear potential (PLP) with new hydrogen bonding and electrostatics terms included. The three-dimensional Crystal Structure of Human Tcf-4/beta-Catenin Complex (PDB ID: 1JPW) has been determined by X-ray crystallography retrieved from Protein Data Bank. During the protein preparation the PDB file has been loaded in the MVD Workspace and only A chain has selected as target of interest. The water Molecule (848) and other Active Ligands has been removed before running the Docking process. The binding cavity was set at three different sites *i.e.* Cavity I at X: 103.918, Y:-28.871, Z: -15.907.

Molecular docking is a method which predicts the preferred orientation of one molecule to a second when bound to each other to form a stable complex. Knowledge of the preferred orientation in turn may be used to predict the strength of association or binding affinity between two molecules using for example scoring functions.

Docking is frequently used to predict the binding orientation of small molecule drug candidates to their protein targets in order to in turn predict the affinity and activity of the small molecule. Hence docking plays an important role in the rational design of drugs. In this work the Docking has been performed at MVD where RMSD threshold for multiple cluster poses was set at < 2.00 Å. The docking algorithm was set at maximum iteration of 1500 with a simplex evolution population size of 50 and a minimum of 10 runs. The post docking analysis has been performed according to Rerank Score.

Pharmacophore Generation

In modern computational chemistry, pharmacophores are used to define the essential features of one or more molecules with the same biological activity. Typical pharmacophore features are for where a molecule is hydrophobic, aromatic, a hydrogen bond acceptor, a hydrogen bond donor, a cation, or an anion. The features need to match different chemical groups with similar properties, in order to identify novel ligands. Ligands receptor interactions are typically "polar positive", "polar negative" or "hydrophobic". A well-defined pharmacophore model includes both hydrophobic volumes and hydrogen bond vectors.

PharmaGist Server

PharmaGist is a freely available web server for pharmacophore detection. The employed method is ligand based. It does not require the structure of the target receptor. Instead, the input is a set of structures of drug-like molecules that are known to bind to the receptor. The main innovation of this approach is that the flexibility of the input ligands is handled explicitly and in deterministic manner within the alignment process. In this current work 5 molecule (Best Re-Rank Score) has been submitted to Pharma Gistand a pivot (Lead I) molecule has been generated.

Results

Results of Protein Parameters and Cavity Detection

Results shown in Table 2.1.

Results of Physiochemical Property and Calculation

Results shown in Table 2.2.

Results of Bioactivity, Toxicity Energy Minimization

Results shown in Table 2.3.

Docking Result with Best HBond Interaction between Lead and Target

Results shown in Table 2.4.

Result of Pharmacophore Generation

Results shown in Table 2.5.

Discussion and Conclusion

Computational molecular docking provides an efficient and innovative apporaches to examine small molecule and protein interaction. Here we are trying to find a novel inhibitor of β-catenin from a group of bioactive anticancer phytochemicals of Indian origin. In this work the ligand based virtual screening has been carried out and four lead (a pivot molecule) compounds have proposed on the basis of Rerank Score (MVD Score). The best lead molecule shows Rerank Score(RS) of -31.229 followed by Lead II (RS: -22.659) and Lead III (RS:4.515). The predicted Scores of MolDock and HBond has been described in Table 2.4.

The Biocativity and Toxicity was predicted for all the compounds in PASS (Prediction of Activity Spectra for Substances). PASS can predict 2000 kinds of biological activity with the mean prediction accuracy of about 87 per cent. In this case only three significant toxic effects were predicted *i.e.* Carcinogenic (A carcinogen is any substance, radionuclide or radiation, that is an agent directly involved in the exacerbation of cancer or in the increase of its propagation), Embryotoxic (Toxicology Adverse effects on the embryo due to a substance that enters the maternal system and crosses the placental barrier; the effects of the substance may be expressed as embryonic death or abnormal development of one or more body

Table 2.1: Results of Protein Cavity Detection

PDB ID	1JPW (Crystal Structure of a Human Tcf-4/ beta-Catenin Complex)
Classification	Cell Adhesion
Structure Weight	192513.59
Molecule	BETA-CATENIN (A,B,C)
Polymer	1 (A,B,C)
Type	Polypeptide(L)
Length	540 AA (Chain A)
Chains	A, B, C, D, E, and F
Resolution	2.50 Å(X-RAY DIFFRACTION)

Protein Cavity Detection

Cavities	Volume	Surface	Coordinates (Position) Z	Y	Z
Cavity I	31.744 Å³	131.833 Å²	103.918	-28.871	-15.907
Cavity II	30.208 Å³	117.76 Å²	105.297	16.244	30.761
Cavity III	27.136 Å³	104.96 Å²	112.781	0.804	42.152

Chain A (Crystal Structure of a Human Tcf-4/β-Catenin Complex) in fasta

```
>1JPW:A|PDBID|CHAIN|SEQUENCE|MGSHAVVNLINYQDDAELATRAIPELT
KLLNDEDQVVVNKAAVMVHQLSKKEASRHAIMRSPQMVSAIVRTMQNTNDV
ETARCTAGTLHNLSHHREGLLAIFKSGGIPALVKMLGSPVDSVLFYAITTLHNL
LLHQEGAKMAVRLAGGLQKMVALLNKTNVKFLAITTDCLQILAYGNQESKLI
ILASGGPQALVNIMRTYTYEKLLWTTSRVLKVLSVCSSNKPAIVEAGGMQALG
LHLTDPSQRLVQNCLWTLRNLSDAATKQEGMEGLLGTLVQLLGSDDINVVT
CAAGILSNLTCNNYKNKMMVCQVGGIEALVRTVLRAGDREDITEPAICALRHL
TSRHQEAEMAQNAVRLHYGLPVVKLLHPPSHWPLIKATVGLIRNLALCPAN
HAPLREQGAIPRLVQLLVRAHQDTQRRTSMGGTQQQFVEGVRMEEIVEG
CTGALHILARDVHNRIVIRGLNTIPLFVQLLYSPIENIQRVAAGVLCELAQDKEA
AEAIEAEGATAPLTELLHSRNEGVATYAAAVLFRMSEDKPQDY
```

Table 2.2: Results of Physio-chemical Property of Best 3 Proposed Leads Calculation

Sl.No.	IUPAC Name	Chemical Formula	HBA	HBD	MW	ClogP	NRB	Groups
Lead I	4-(10-formyl-2,8,9-trihydroxy-7-isopropyl-4-methylbicycle [4.4.0] deca-1,3,5,7,9- pentaen-3-yl)-5,8,9-trihydroxy-10-isopropyl-3-methyl-bicyclo[4.4.0]deca-1,3,5,7,9-pentaene-7-carbaldehyde	$C_{30}H_{30}O_8$	8	5	518.941	6.3	5	Aldehyde Alcohol (Hydroxyl)
Lead II	2,6,6-trimethyl-cyclohexa-1,3-dienecarbaldehyde	$C_{10}H_{14}O$	1	0	150.104	1.28	1	NP
Lead III	NP	$C_{15}H_{18}N_4O_5$	6	5	334.127	-3.51	4	Amine Ether Aziridine

NB: HBA: No of Hydrogen Bond Acceptor, HBD: No of Hydrogen Bond Donor, MW: Molecular Weight (Da), NRB: No. of Rotatable Bond, NP : Not Predicted.

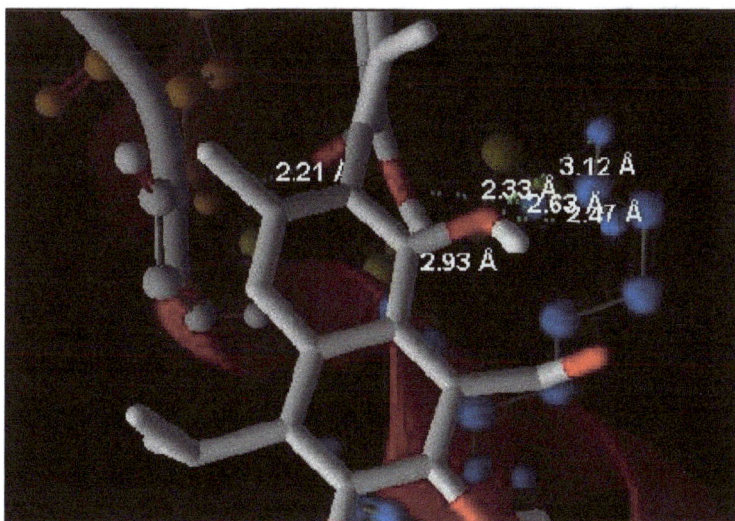

Figure 2.1: H-Bond Interaction Between Lead I with Target Protein

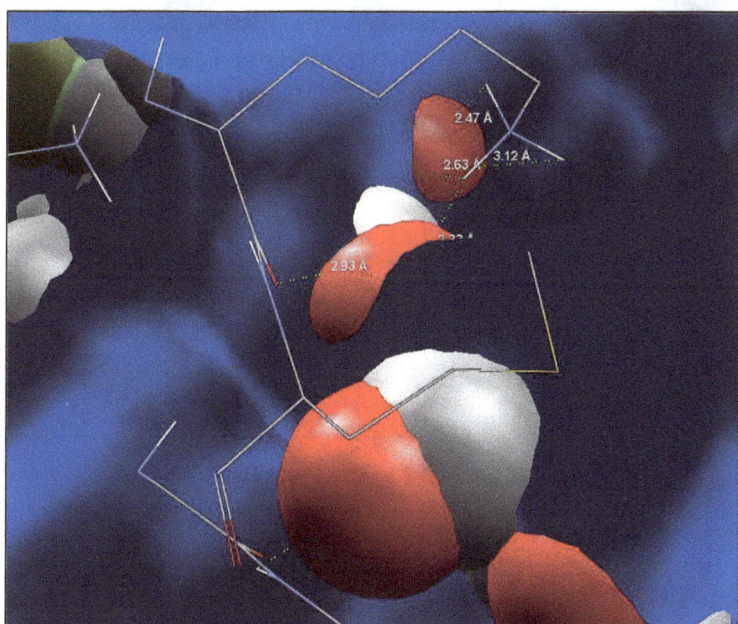

Figure 2.2: Interaction between Lead I with Target Protein (Surface)

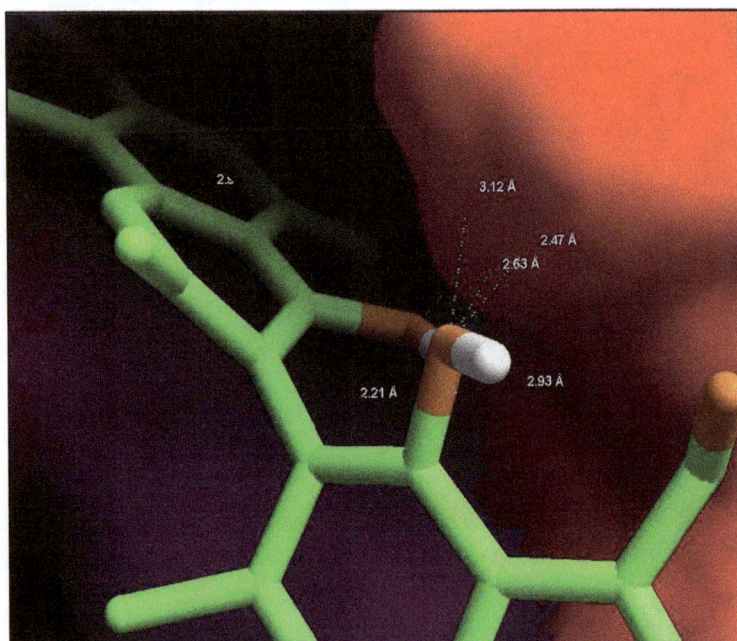

Figure 2.3: Interaction between Lead I with Target Protein (Hydrophobicity)

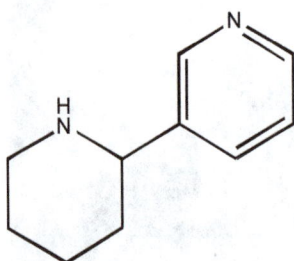

3-(piperidin-2-yl)pyridine

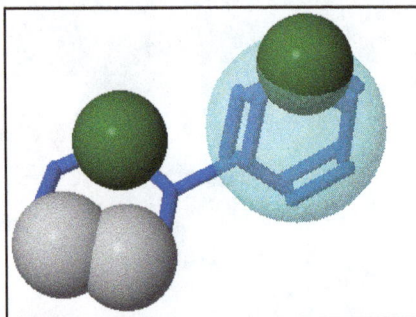

Figures 2.4 and 2.5: The Pivot Molecule

systems, and can be deleterious to maternal health) and Teratogen (A drug, or radiation, that can cause malformations in an embryo or fetus). The 21 Ligands (compounds) was found to be non-toxic and the physiochemical properties has calculated against these compounds, where Lead III (HBA=6, HBD=5, MW= 334.127, *ClogP*

Table 2.3: Results of Bioactivity, Toxicity and Forcefield Prediction of Best 3 Lead Compounds

Sl.No.	Leads with 2D Structure	3D Structure	Bioactivity and Toxicity Prediction				Force Field Calculation (Total Energy in kcal/mol)	
			Drug-Likeness	Activities	Pa	Pi		
Lead I			0.765	1. Endothelin converting enzyme inhibitor	0.160	0.064	Stretch	9.134
							Bend	32.545
							Stretch-Bend	0.635
							Torsion	22.319
							Non-1,4 VDW	28.578
							1,4 VDW	28.578
							Total Energy	**99.708**
Lead II			0.434	1. Antidiuretic hormone antagonist	0.070	0.025	Stretch	1.471
							Bend	2.961
							Stretch-Bend	0.144
				2. Vasopressin II antagonist	0.062	0.059	Torsion	0.528
							Non-1,4 VDW	1.346
							1,4 VDW	8.088
							Total Energy	**13.546**

Contd...

Table 2.3–Contd...

SI.No.	Leads with 2D Structure	3D Structure	Bioactivity and Toxicity Prediction				Force Field Calculation (Total Energy in kcal/mol)
			Drug-Likeness	Activities	Pa	Pi	
Lead III			0.994	Toxicity Not Predicted			Stretch 5.239 Bend 142.155 Stretch-Bend -5.163 Torsion 4.746 Non-1,4 VDW -2.147 1,4 VDW 10.630 **Total Energy 157.025**

Table 2.4: Hydrogen Bond Interaction between Best Three Ligand with Target Protein

Ligands							Protein (Chain A): Cavity I()		
Proposed Ligand	MolDock Score	Rerank Score	HBond	Atom Name	Atom ID	Distance of H bond (Å<3.5)	Atom Name	Atom ID	Amino Acid
Lead I	-42.663	-31.229	-10.191	O(8)	36	2.21 Å	H(1)	4543	Ser246
				O(8)	25	2.33 Å	H(1)	4518	Lys242
				O(8)	18	2.47 Å	H(1)	4519	Lys242
Lead II	-33.245	-22.659	-0.0117	O(8)	10	3.18 Å	H(1)	4246	Asp207
				O(8)	10	3.49 Å	H(1)	4272	Ala211
Lead III	-38.879	4.515	0.635	H(1)	41	0.84 Å	O(8)	390	Met202
				H(1)	35	2.23 Å	O(8)	415	Thr205
				O(8)	18	2.33 Å	H(1)	4517	Lys242

Table 2.5: Pharmacophore Features of the Pivot Molecule

Score	Features	Spatial Features	Aromatic	Hydrophobic	Donors	Acceptors	Negatives
9.341	6	6	1	1	1	3	0

=-3.51 and NRB=4) shows better result in comparison with other two lead compounds. The drug-likeness of Lead I molecule was predicted as 76.5 per cent followed by Lead III (99.4 per cent) and Lead II (43.4 per cent) as in Table 2.3. During the Forcefield/Energy Minimization Calculation (Table 2.3) Stretch, Bend, Stretch-Bend, Torsion, Non-1, 4 VDW, 1, 4 VDW and Total Energy in kcal/mol were predicted for each lead molecule, where Lead II shows better Energy (Total) Value of 13.54 kcal/mol. In order to perform docking the Protein (Target) Cavities has been predicted. The best cavity was detected at X: 103.918, Y: -28.871 and Z: -15.907 in MVD workspace.

The Docking results were analyzed on the basis of rerank Score. The rerank Score is a linear combination of E-inter (steric, Van der Waals, hydrogen bonding, electrostatics) between the ligand and protein, and E-intra (torsion, sp2-sp2, hydrogen bonding, Vander Waals, electrostatics) of the ligand weighted by pre-defined coefficients. It was notice that Lead I (Table 2.4) shows HBond interaction (Hydrogen Bond Interaction) with protein at 2.21 Å (Ser246), 2.33Å (Lys 242) and 2.47 Å (Lys242), whereas Lead III (3 HBond interaction) resulted close contact at 0.84 Å (Met202). The Lead II annotated very poor HBond interaction (2 interactions). The Pharmocophore model has been generated and described in Table 2.5 and Figures 2.4 and 2.5.

The ligand based virtual screening result endeavors potential of these 4 lead compounds including pivot molecule *i.e.* 3-(piperidin-2-yl)pyridine (Score 9.341), would open up new avenues for designing of potent Human β-catenin inhibitors if synthesized and tested in animal models.

Acknowledgement

Dr. R.K. Singh, Coordinator, Bioinformatics Infrastructure Facility, Rajiv Gandhi University, Itanagar, Arunachal Pradesh, Dr. M.J. Bordoloi, Scientist EII, Natural Product Chemistry Division, North-East Institute of Science and Technology (CSIR), Dr. R.L. Bezbaruah, Scientist F, Biotechnology Division, North-East Institute of Science and Technology (CSIR), Jorhat, Assam.

References

Botrugno, O.A., Fayard, E. and Annicotte, J.S. *et al.*, 2004. Synergy between LRH-1 and h-catenin induces G1 cyclin-mediated cell proliferation. *Mol. Cell.,* 15: 499–509.

Clevers, H., 2006. Wnt/h-catenin signaling in development and disease. *Cell,* 127: 469–480.

Cragg, G.M. and Newman, D.J., 2001. Natural product drug discovery in the next millennium. *Pharmaceut. Biol.,* 39: 8–17.

Farnsworth, N.R., Akerele, O., Bingel, A.S., Soejarto, D.D. and Guo, Z., 1985. Medicinal plants in therapy. *Bull WHO,* 63, 965–981. (ref of anticancer-04).

Goodsell, D.S. and Olson, A.J., 1990. *Proteins: Str. Func. and Genet.,* 8: 195-202. "Automated Docking of Substrates to Proteins by Simulated Annealing".

Haber, D. and Harlow, E., 1997. Tumour-suppressor genes: Evolving definitions in the genomic age. *Nature Genet.,* 16: 320-322.

Holt, G.A. and Chandra, A., 2002. Herbs in the modern healthcare environment: An overview of uses, legalities, and the role of the healthcare professional. *Clin. Res. Regulatory Affairs (USA),* 19: 83–107.

Jansson, E.A., Are, A., Greicius, G. *et al.,* 2005. The Wnt/h-catenin signaling pathway targets PPARg activity in colon cancer cells. *Proc. Natl. Acad. Sci., USA,* 102: 1460–1465.

Krishnaswamy, K., 1996. Indian functional foods: Role in prevention of cancer. *Nutr. Rev.,* 54: S127.

Kuo-Hsiung, L., 1999. Anticancer drug design based on plant-derived natural products (Review). *J. Biomed. Sci.,* 6: 236-250.

Lalramnghinglova, J.H., 1999. Prospects of ethno-medicinal plants of Mizoram in the new millennium. In: *Symposium on Science and Technology for Mizoram in the 21st Century,* (Eds.) Laltanpuia and R.K. Lallianthanga, pp. 119–129.

Lipinski, C.A., Lombardo, F., Dominy, B.W. and Feeney, P.J., 1997. Experimental and computational approaches to estimate solubility and permeability in drug discovery and development settings. *Adv. Drug Del. Rev.,* 23: 3–25.

Moon, R.T., Kohn, A.D., De Ferrari, G.V. and Kaykas, A., 2004. WNT and β-catenin signalling: diseases and therapies. *Nat. Rev. Genet.,* 5: 691–701.

Nath, N., Kashfi, K., Chen, J. and Rigas, B., 2003. Nitric oxide-donating aspirin inhibits h-catenin/Tcell factor (TCF) signaling in SW480 colon cancer cells by disrupting the nuclear h-catenin-TCF association. *Proc. Natl. Acad. Sci., USA,* 100: 12584–12589.

Prasad, S.B., 2005. Cancer and anticancer drug, cisplatin. In: *Cancer: A Cytogenetic and Molecular Approach,* (Ed.) V. Rai. Bioved Research Society, Allahabad, pp. 159–176.

Rosangkima, G. and Prasad, S.B., 2004. Antitumour activity of some plants from Meghalaya and Mizoram against murine ascites Dalton's lymphoma. *Indian J. Exp. Biol.,* 42: 981–988.

Singh, R., Artaza, J.N. and Taylor, W.E. *et al.,* 2006. Testosterone inhibits adipogenic differentiation in 3T3-1 cells: Nuclear translocation of androgen receptor complex with h-catenin and T-cell factor 4 may bypass canonical Wnt signaling to down-regulate adipogenic transcription factors. *Endocrinology,* 147: 141–154.

Sllchenmyer, W.J. and Von Hoff, D.D., 1991. Taxol: A new and effective anticancer drug. *Anticancer Drugs,* 2: 519–530.

Chapter 3

Biodiversity, Traditional Knowledge, IPR and Rationalization of Traditional Phytomedicine in the Context of North East India

Hui Tag*, P. Kalita and A.K. Das

ABSTRACT

Present paper discusses biodiversity potential, traditional knowledge (TK), scope for rationalizing traditional phytomedicine of Northeast India with special reference to current IPR regime, and its possible implication on TK based phytomedicine sector. Comparative literature studies revealed that NE region of India (NEI) is rich in both biodiversity and TK heritage but traditional phytomedicinal products needs rationalization at various level of pharmacognosy research. Such wide gap in R&D sector confers ample scope for the bioprospecting work on numerous species of medicinal with deep cultural history of usage available in the NEI. However, the majority of university and research institute of the region

---

*   Corresponding author: E-mail: huitag2008rgu@gmail.com

is suffering from poor R&D infrastructure coupled with acute shortage of dynamic manpower to motivate the young researchers. Capacity building in traditional phytomedicine research sector, standardisation and quality control of traditional phyto-herbal products is the major thrust area where government should play a big role by patronizing academic and research institutes of NEI. On the other side of the coin, though the region is rich in biodiversity and TK heritage but recognizing IP Rights of TK holders by following CBD guidelines are yet to take its effect, hence there is an urgent need for IPR education at all level of people. Recognizing IP Right of the TK holders before and after patenting of products by researchers can minimize current rate of biopiracy and genetic erosion, boost the morale of the community as well as regional economy of NEI.

Keywords: *Biodiversity, Traditional phytomedicine, Rationalization, Traditional knowledge, IPR, NE India.*

Introduction

The States of Northeast India fall within East Himalaya and Indo-Burma Biodiversity Hotspots with rich relics of broad leaved forest coverage and high diversity of rare and endemic flora and fauna (Myers *et al.,* 2000). The region is also a veritable home to about 8 million tribal people clubbed within 150 tribal communities with diverse ethnocultural heritage. About 1953 ethno-medicinal plants have been reported from the NE region of which 1400 species are used as both food and ethno-medicinal sources (Dutta and Dutta, 2005; Albert and Kuldip, 2006). Among the state of Northeast India, Arunachal Pradesh (83,743 sq.km), which formed a part of Arunachal Himalaya (AH) is not only the largest but also a richest forest coverage (68 per cent) with rich heritage of flora and faunal diversity of both ethnocultural and ecological significance that falls within Himalaya Biodiversity Hotspots (Hegde, 2002). This high diversity is due to the presence of all climatic zones ranging from tropical to snow-clad alpine mountains (Kaul and Haridasan, 1987). Almost 33 per cent of higher plants and 52 per cent of orchid species of India are found in the forest of Arunachal Himalaya (Mao and Hyniewta, 2000; Hegde 2002). The AH region has 26 major tribes and 110 subtribes with rich tradition of age-old indigenous knowledge system (Tag and Das, 2004). However, in the current world order, liberalization of global trade policies and other economic activity reforms evolving with emergence of the UN Conventions on Biological Diversity (CBD, 1992), and the World Trade Organization (WTO) needs deeper study and understanding especially in the light of latest path breaking achievement in science and technology. It is pertinent to note that the Northern Developed Nations-also called scientifically advanced and legally skilled nations continue to take undue advantage of the bioresources and traditional knowledge of the Southern Nations called third world countries (TWC), primarily a poverty ridden and developing nations (Puspangadan and Nair, 2005). States of Northeast India is still in nascent state especially in the field of biodiversity, traditional knowledge and IPR related research and awareness. Such nascence states is not only prone to resources exploitation and misappropriation of its rich traditional knowledge heritage by legally skilled nations but also susceptible to remain underdeveloped if its TK related bioresource potential are not tapped properly in scientific dimension. Ethnobotanical literatures are available in plenty based on preliminary field work from seven states of NE India but few reliable literatures on quantitative disease specific ethno-medico-botany based on rigor methodology are available to date (Nima *et al.,* 2009, Tag *et al.,* 2009). In current decades, Traditional Phytomedicine (TPM) are attracting global attention within the context of health care provision and health sector reform but need rationalization in order to ensure health safety and to prevent possible misappropriation of its associated TK. Rationalization of traditional phytomedicine bears immense scope in the context of Northeast India as the region is treasurer house of traditional medicinal plants used across different tribes and races and its efficacy has been proven through odyssey of time in folk medical system. However, to tape richness of potential plants in sustainable approach and to prevent piracy of associated TK, systematic and rigorous pharmacognostic investigation of potential species using modern tools and techniques, strict implementation of IPR and Patent Law is urgently required. Against these backdrops, present paper aims to highlight hidden biodiversity potential, traditional knowledge and related IPR issues, and also highlights major scopes in the realm of traditional knowledge based phytomedicine and safety issues in the context of North East India.

Biodiversity and Traditional Knowledge: Global Scenario

Biodiversity is the invaluable wealth on which human livelihood system is entirely based. It is evident through the history that the economic activity of humankind depends directly or indirectly from the biological resources for its sustenance since time immemorial. Biodiversity or biological diversity is defined as total sum of all living form including microbes. On the other hand it is defined as variety and variability of living organisms on the planet Earth which form the basis for sustainable economic development. Till date, Systematic Botanist and Systematic Zoologist together have documented 1.4 million species of living organism which includes microbes, fungi, plants and animals. Not satisfied with present miraculous achievement done within 300 years of Systematic Biology Research, the optimistic Biological Community are still looking forward with close estimation that atleast 6 million species of earth biosphere are yet to be documented in scientific literature in the next 50 years (Myers *et al.,* 2000). However, apprehension is inevitable as the current rapid depletion of global forest coverage at the rate of 350 million hectares between the years 1990-1995 has been witnessed whereas as per Millennium Ecosystem Report (2008) we have in our custody only 3500 million hectares of forest coverage and is predicted that 50 per cent reduction in global forest coverage will take place by the end of year 2050. The present trends of rapidly depleting world bioresources across the globe due to anthropogenic threat has added much to the apprehension of systematic biologist that many rare, endangered as well as endemic species which are yet to be recorded in scientific literature would disappear faster than the speed in which present uninterrupted speed of documentation activity happening within the scientific community across the world. Greatest part of the global biodiversity (species diversity) is found in animal and microorganisms. Over two third of the estimated 300,000 species of higher plants in the world occurs in the tropical forest of south America, Africa, Madagascar and tropical Asia including New Guinea and tropical Australia. Most of the countries located in these regions are interestingly the TWC blessed with almost all known type of topographic and climatic conditions ranging from tropical to temperate and alpine zones while the tropical rainforest alone harbours 50 per cent of Earth's Species Diversity. Around 1,26,756 species of plants, animals, fungi and microorganisms has been identified and classified from Indian subcontinent. Around 70 per cent of species diversity has been reported from Indian two biodiversity hotspots- The Western Ghat and Eastern Himalayas and these hotspots are guarded by tribal communities and their age-old traditional knowledge system (Karthikian, 2000, Joshi and Joshi, 2004). Interestingly the rich biodiversity of India is equally matched with rich cultural diversity and a unique wealth of traditional knowledge system, developed, preserved and practiced by millions of indigenous people living in the rural localities and forest areas. Biodiversity performs two most important functions. Firstly it regulates and maintains the stability of climate, water regime, soil fertility, and quality of air and overall health of the life support of earth's ecosystem. Secondly, biodiversity is the source from which human race derive food, fodder, fuel, fiber, shelter, medicine and raw materials for meeting other multifarious human and industrial needs, therefore, the biodiversity is the biological capital of our planet and it forms the foundation upon which human civilization is built and would prosper further. Fortunately the third world nations are not only rich in agricultural and natural resources (biodiversity) but also equally rich in TK system. The megadiverse country such as India and China has more than 8000-5000 years of continuous cultural history and TK. In present scenario, the TK is understood as a community based system of knowledge that has developed, preserved and maintained over many generations by the local communities through their continuous interactions, observations, experimentations with surrounding environment. TK is unique to a given culture or society and is developed as a result of co-evolution and coexistence of both the indigenous culture and traditional practices of resources use and ecosystem management. TK in general term refers to collective knowledge, beliefs and practices of indigenous people on sustainable use and management of their ambient resources. Through years of observations and analysis, trial and error process of experimentations, the traditional communities have been able to identify useful as well as harmful elements of their ambient flora and fauna. When talk of Intellectual Property Right (IPR), it has direct correlation with TK. In laymen concept, IPR is understood as intellect or knowledge of individual or community in the form of innovation, creation, healing techniques, arts and craft knowledge, folklore etc. which is treated as property of that individual/community who have developed in a particular period of time or developed, preserved or possessed over time. On the other hand Intellectual Property Right is defined as ownership right of the individual/community over their creation and innovation out of their creative thought. Here IPR could be both traditional and non-traditional type. The traditional knowledge acquired through the ages has always remained as part of tribal life, culture, traditions beliefs, folklore, arts, music, dances etc are treated as the intellectual property of particular tribal community who possessed that valuable knowledge and therefore each tribal or local community have the ownership rights over their biodiversity-based, arts-based and folklore-based TK because TK of every culture is unique and location specific (Puspangadan, 2002). The TK covers broad spectrum of the indigenous people's traditional life and culture, arts music, architecture, agriculture, medicine, engineering and host of other spheres of human activity. Therefore, TK can be of direct or indirect benefit to society as it is often developed in part as an intellectual response to the necessities of their life. Protection and maintenance of TK of local and indigenous communities through IPR regulations is vital for their socio-economic well being, intellectual and cultural vitality through which sustainable development can be achieved. The accumulated wisdom, knowledge, beliefs and practices embodied in the TK system were handed down from generation by unbroken chain of tradition and culture. Such tradition is still living in many parts of the third world nations while Indian and Chinese tradition of science, agricultural and medicinal practices are still considered oldest and time tested which need strict IPR regulations to protect the ownership rights of the indigenous community at individual or community level. Earlier, the traditional society in India as well as in other third world nations have always considered the biodiversity (natural resources) and its associated TK system developed by them as commonly owned properties (Common Intellectual Property) to be guarded and shared by all but never to be commodified for the purpose of selling or marketing. It was with the coming of the European that the process of commodification and trading of Earth's Bioresources and its associated TK/IPR started during first half of 16th Century in Europe with commercial tactics. Such bioresource commercialization tactics, no doubt, has done many good to the indigenous communities of TW countries by bringing them to the societal mainstream but its negative impact has been felt throughout Asia, Africa, Oceania and Latin American nations in the form of colonization, resource exploitation, habitat destruction, cultural marginalization, religious conversion, and trade monopoly that has done much to the detriment of the local socio-cultural fabrics, TK-based IPR and traditional ecological environment of the indigenous people of third world nations (Mashelkar, 2001).

Relevance of Indian Biological Diversity Act (2002)

In line with Convention on Biological Diversity Act (1992) for protection of biodiversity and its related TK, Indian Government has finally enacted Biological Diversity Act (2002) which aims at protection and promotion of the Country's rich biodiversity heritage for the sustainable existence of humanity and environment. Under the provision of BDA (2002), the national, state and local biodiversity boards/committee are entrusted to oversee and implements benefit sharing mechanisms, documentation of bioresources and traditional knowledge, material transfer, access agreements on genetic resources and technology etc. The Acts stipulated norms for access to biological resources and traditional knowledge based on three ways: 1. Access to biological resources and traditional knowledge to foreign citizens, companies and NRI based on prior approval of National Biodiversity Authority (NBA), Chennai (Sec. 3 of the Act); 2. Access to Indian citizens, companies, associations and other organization registered in India on the basis of prior information to the State Biodiversity Board (SBB) concern (Sec. 7 of the Act); 3. Exemption of prior approval or intimation for local people and communities, including growers and cultivators of biodiversity, and *vaids*, *hakims*, and herbal practitioners, who have been practicing indigenous medicines for human and animal welfare (Sec 7 of the Act). However, it is difficult for the policy makers and even the concerned indigenous community that how much TK related biodiversity are nurturing within their locality or the region that need top priority for conservation action plan. Therefore, the National Biodiversity Authority (NBA) has setup guidelines for State Biodiversity Board (SBB) to maintain community biodiversity register so that biodiversity wealth nurture by the community and their associated TK are kept in record, so that concern indigenous community could act as watchdog to protect the resource and associated TK from infringement. It also confers comprehensive rights to indigenous community over their local biodiversity resources which enable the community to conserve their own biological resource in traditional approach in their community forest, agricultural lands and home garden for sustainable livelihood. The Indian sub-continent is multilingual and multiethnic region and each indigenous community (ca 570) have their own biodiversity based traditional knowledge and cultural heritage. Most of their primary and secondary economic activities such as agriculture, fishing and hunting, medicinal items, handicraft design, furniture, rural utensil, house construction, ritual items are forest products based such as bamboo, wood, medicinal herbs and animal products. Each tribal community has traditional knowledge which is quite unique and sacred to them, which need to be preserved and protected under Indian Forest Law, Forest Right Acts (2006), IPR Laws and Indian Patent Law from misappropriation. If we closely examine Indian Forest Right Act (2006), though the act confers sovereign rights to tribal community to use forest products including medicinal plants from government reserve forest but beurocratic interference and permit raj is preventing these tribal communities from commercializing such forest resources for their economic benefit. If we closely examine Indian Patent Law, some of the techniques used by the tribes of Indian sub-continent are patentable such as herbal formulation techniques to cure particular ailments, handicraft design and decoration, formulated plant origin herbicides and pesticides, fishing and trapping techniques, bird catching techniques etc.

Richness of Biodiversity or Biocultural Diversity in NE India

The North East India, also known as the land of the eight sisters, comprises the States of Arunachal Pradesh, Assam, Manipur, Meghalaya, Mizoram, Nagaland, Sikkim and Tripura, which collectively account for about 8 per cent of the country's geographical area, constitutes 4 per cent of total Indian population. The region is known for its ethnic, linguistic, cultural, religious and physiographical diversity. The region is abode of around 150 major tribes and 300 communities which represent around 31 per cent of the total 475 tribal community found in India.WWF has identified entire NE region of India as priority Global 200 Ecoregion while IUCN and Conservation International has included entire North East India as one of the top 12th Global Biodiversity Hotspots in

Eastern Himalayas in the year 2000 by virtue of its rich cultural and biological diversity (Chatterjee *et al.*, 2006). As per Forest Survey of India Report (2005), the NE region has total forest coverage of 173316 Km^2 which represent 66.1 per cent of its total geographical area of NE India, which is far above national average, and ideal habitat for the proliferation of faunal biota. Its unique location at the confluence of the Indo-Malayan, Indo-Chinese and Indian biogeographical regions coupled with its diverse physiography has generated a profusion of habitats, which harbours diverse biota with high degree of endemism (Ashok, 2000). Due to the high degree of endemism in floristic and faunal elements, the Indo-Myanmar region has been cited as centre for origin of flowering plants (Myers, 1988). The entire Eastern Himalayan region is estimated to have more than 13,500 species of flowering and non-flowering plants wealth which represents around 33 per cent of Indian flora and represent 2.3 per cent of global plant diversity, of which 7000 species are endemic to the region. Furthermore, the region harbours 800 species of critically endangered plants out of 1500 species of endangered plant species identified in India. Arunachal Pradesh alone with more than 5000 species of flowering plants is also considered as richest treasure house of both plant and animal species including microbial diversity of ecological, medicinal and cultural importance (Karthikiyan, 2000, Mao and Hyniewta, 2000). The region is rich enough in Orchids species which represents 55 per cent of total orchid species diversity in India, around 90 species of Bamboo which is highest in any other states in India and just next only to China. Around 1400 species of medicinal plants are found in the NE region which represents around 70 per cent of medicinal flora found in India (Hegde, 2000, Dutta and Dutta, 2005). It is interesting to see that the NE region of India has been identified by Indian Council of Agricultural Research (ICAR) as centre for origin of rice germplasm while National Beurea of Plant Genetic Resources (NBPGR) has identified the NE India as Hindustan Centre for the origin of cultivated crop species on the basis of availability of numerous indigenous cultivars and varieties of fruits, vegetable and cereal crops natured by more than 300 tribal communities and clans in their traditional *jhum* land and home garden. Around 60 land races of cultivated cereal crops are found in this region which is highest in any other country or region of the world. The NE region is equally rich in animal diversity (FAO, 2004).

Current figure tells that 2185 species of vertebrate are available in NE India of which 528 species are endemic to this region. Mammalian diversity constitutes around 329 species out of which 73 species are endemic to the region; reptilian diversity represents 484 species and 202 species are amphibians. The avian fauna is represented by 1170 species, of which 140 species are endemic to this region and 21 species are migratory (Joshi and Joshi, 2004). Interestingly, ethnozoological studies revealed that faunal species are often used as major components in formulation of traditional phytotherapeutic system, and majority of them being hunted for food, medicine and cultural reasons (Borang, 1996). The rich biodiversity and cultural diversity of region could come under threat if the policy makers fail to check the current ever increasing trends of demographic figure, and also the trends of unsustainable resource exploitation in the region. It is very difficult to check and the reason is multifarious. The 2001 census has put total population of NE India at 39015582 persons where Assam lead the highest density of population with 320 persons per km^2 which is followed by Tripura with 304 persons per Km^2. Here, Arunachal Pradesh stands lowest in population density graph with only 13 persons per Km^2 though it is rated as largest state in NE India in terms of geographical area and richest state in terms of biodiversity wealth and hydropower potential. Looking back to the historical background, successive legal and administrative decisions taken by the British between 1874 and 1935 gave the areas of the Northeast their distinct identity which indirectly contributed significantly in conserving biodiversity and traditional ethnic culture. To date, the states of NE India have been able to retain its significant proportion of biodiversity and the rate of genetic erosion is not too pathetic when compared to other part of the country due to its long years of isolation, remoteness and inaccessible physiographic terrain. However demographic change and upcoming developmental projects in region is putting pressure on its forest and biodiversity wealth and many rare and endangered medicinal plant species of high commercial values are disappearing fast or either pirated by foreign national un-noticed. The magnitude of development impact and infringement of community rights over local biodiversity and associated TK is being felt in all parts of NE region which needs urgent attention from the policy makers and common citizens.

Traditional Phytomedicine in NE India: Prospects and Constraints

As per WHO definition, traditional phytomedicine is purely a plant based medicinal products used for curing various diseases and ailments by applying traditional knowledge. The investigation of natural products as source of novel human therapeutics reached its peak in the western pharmaceutical industry during 1970-1980, which resulted in a pharmaceutical landscapes heavily influenced by non-synthetic molecules. Of the 877 small molecules NCE introduced between 1981 and 2002, roughly half (49 per cent) were natural product majority being botanical origin, semi-synthetic NP analogues or synthetic compounds based on NP. Current phytomedicine literature revealed that around 80 per cent of population in developing nations relay on traditional herbal medicine. Almost 75 per cent of phytomedicine developed by modern industries are ethnobotanical knowledge based while 25 per cent of modern allopathic drugs in developed nations are plant based (WHO, 2005). This implies that traditional phytomedicine still plays a bigger role in both developing and developed nations in human healthcare sector but also leaving ample scope for standardisation and strict regulation under drug and cosmetic acts of respective member nations across the globe. A case study on characteristics of traditional phytomedicine has revealed several merit points. The first merit point is its reliability as the system is proven by centuries old traditional cultural knowledge, and other merits are the lesser side effect, cheap and affordable for the poor, purely a botanical origin (Fabricant and Farnworth, 2002, Patwardhan, 2007). Ethnobotanical research in 21st century has unfolded hidden potential of medicinal plants from NE India. Of 2500 ethnobotanical species, 1953 species have been identified as medicinal and such diversity confers diverse scope for rationalization at phytochemistry, molecular, pharmacobiotechnology, tissue culture and pharmacogenomic level. Ethnobotanically guided case studies on the traditional phytomedicine system in NE India shows that phytomedicine are mostly rural based medicine where drug formulation are usually conventional type, admixture of more than single plant parts, compound formulation is disease specific, strong traditional cultural knowledge based, often do miracle in curing worrisome ailment like cancer, HIV, and brain related disorders.

Dose formulation of phytomedicine is approximate based on empirical knowledge of TK holders, but need rationalization and validation study in modern laboratory (Jain, 1999; Tag *et al.,* 2008).

Currently, there are several prospects for validation of phytomedicine through ethnopharmacology research in NE India. In view of the vast potential of medicinal plants species with deep rooted cultural history available in NE India, capacity building in the area of core and applied pharmacognosy, phytochemistry and molecular research is urgently needed in lead institutes and Universities of the NE region to rationalize traditional claims of phytomedicine, and to ensure standardization, safety and efficacy. Traditional phytomedicine used by the ethnic communities provides ample scope for rationalization based on specific ailments such as plants with anti-inflammatory activity, wound healing and antioxidant activity, anti-diabetic activity, anti-malarial activity, anti-tumour activity, anti-diarheoal activity, biochemistry and anti-oxidant activities of traditional food plants, plants with central Nervous System activity, blending traditional herbal formulation techniques with modern pharmacological and pharmacogenomic science. After proving efficacy of phytomedicine in animal model experiments as mentioned above, the researchers can further go for isolation and elucidation of compound responsible for curing specific ailments. There are two major approaches to drug discovery that is Common Approaches which includes Chemical Biology Approach, Serendipity and Synthetic, and Combinatorial and Genomic Approaches which often takes about 8-10 years to isolate the novel entity of the plant. However, current innovative approaches based on Traditional Medicine are- Ethnopharmacology Approach, Reverse Pharmacology Approach, Systems Biology Approach, and Personalized Approach which usually takes 4-5 years to isolate the novel entity. Other area of concerned is the safety and quality control of the traditional herbal products as majority of this sector operate without regulatory mechanisms. Government agency should address such unorganized sector by enforce strict guidelines of WHO to ensure standardization and safety through rigorous research in regional natural products laboratory available in the region. However, researchers of NE India may have to face few constraints in rationalizing traditional phytomedicine. The current major bottleneck is that the NE region of India is still drastically lacking with high-tech infrastructure excepting IIT

Gawahati and Tezpur University. In majority of academic institutes, R&D sectors, there is a lack of dynamic and quality human resource. Surprisingly, there is also a lack of Skill Based Training Centre (SBTC) for young researchers in pharmacology, Systematic Botany, Pharmacobotany and molecular science. Such infrastructure lacunae and policy loophole in academic, R&D sectors often discourage the young graduates from moving further towards the height of excellence. Policy constraints in bioresource related patent, IPR that includes inadequate current national and international patent law need to be addressed at the highest level of decision making forum but in the first step, awareness campaign have to be generated from grass root level to enhance quality human capital through academic and research enrichment programme in existing academic and research institutes located in the region.

Biodiversity and Traditional Phytomedicine with Reference to IPR Issues in NE India

The Indian Patent Act (1970) is significant in the sense that it stipulates disclosures of the source and geographical origin of the biological materials in the specification, when used in invention. This issue of disclosure of source of geographical origin is a continuous one that the developing and least developing nations are still pushing ahead with WTO and TRIPS. The new Patent Act of India (2005) includes two important clauses for revocation of a patent on the ground that – i. complete specification does not disclose or wrongly mentions of geographical origin of biological materials used for the invention; ii. the invention so far as claimed in any claims of the complete specification is anticipated having regard to the knowledge, oral or otherwise, available within any local or indigenous community in India or elsewhere. These two provisions ensure protection of the rights of the source of country of a biological materials or traditional knowledge of local or indigenous community, and thereby enabling recognization and reward of source countries and traditional knowledge holders through appropriate benefit sharing mechanism (Jeffrey, 2007). However, global patent scenario reflects that some 95 per cent of bio-resources related patents are hold by developed nations while almost 70 per cent of Earth bioresources and medicinal plant diversity are concentrated in Southern Block which is primarily developing poor tropical nations. This implies that developing nations holds only few (5 per cent) of bioresource related patents despite rich bioresources and associated TK heritage. Such worrisome figure is obviously posing serious challenges to developing nations like India to secure its biodiversity and associated TPM based IPR with diligent and hard work in R&D sector. Nationwide IPR education and social movement is the need of the hour as almost 70 per cent of Indian population are rural based and there is every possibility of plant based TK piracy by northern block nations if grass root level communities are unaware. Current international patent law stipulated that living organism or species are non-patentable but its products such as novel chemical entity, traditional herbal formulation techniques for particular ailments, ethnic recipes, and compound formulations are all patentable (WIPO, 2000, Shiva *et al.,* 2002). Although, the NE region of India with diverse indigenous communities are rich in TK and biodiversity bears immense scope for products commercialization, but the region is drastically lacking in R&D infrastructure, and also lacking with IPR and Patent related education and awareness at grass root level. The IPR and patent education should be incorporated as part of school curriculum design so that students become aware of its significance and utility. Impact of alien culture has been witnessed in NE India where younger generation of NE youths are slowly becoming least concerned to learn and realize the significance of biodiversity based TK of their own locality. Such laxity approach of young generation NE Indian coupled with lack of infrastructure in academic and R&D sector should be matter of serious concern to the Indian patent law makers, national and regional policy makers as well. Scientist and policy makers of NE India must now think some best alternative to stop the piracy of biodiversity and associated TK of NE India. Some of the research of both Indian and foreign origin doing research in NE India region are very secretive. Most of the researchers are connected some way or other with some big multinational company. In recent decades, some of the TK based potential medicinal plants of NE India were collected from the herbalists but bioprospecting and bioactive compound isolation are being carried out in elsewhere advance laboratory of the world. Hundreds of patent are being secured in individual scientist name or name of the concerned laboratory or pharmaceutical company, and papers have been published in high impact international journals but in the acknowledge part, the authors/scientists have mostly failed to disclose the source country and also failed to acknowledging the name of TK holders of particular community. Such cases are more prominent in developed nations where attempt made for IPR infringement by western nations have been challenged by CSIR directorate of India decades ago especially in *Neem, Basmati* and *Haldi* case. IPR infringement is not only done by foreign national but such mistake is also knowingly committed by Indian national where better exposed Indians too are not lagging behind in pirating bioresources related TK of ethnic dominated region of India including NE India without acknowledging the community IPR rights over biodiversity and associated TK. Such acts of few individualistic researchers are the serious violation as per CBD's Prior Informed Consent (PIC) and Access and Benefit Sharing (ABS) guidelines. To abate such incidence, National Innovation Foundation (NIF), a sister organization of DST has shown examples by setting a model format for seeking PIC of the community before accessing their genetic resources and its associated TK. Such IPR movement in research has generated awareness among the tribal communities of Central, Southern and Northern India but such IPR movement is yet to gain its momentum in least developed states of NE India. The NIF PIC and ABS model is worth listening in paper and few success story has been published in some journals and newsletter but the conditions mentioned in their format for seeking PIC from the TK holder has to be critically examined through screening committee (third person observation) so that the TK holders are not cheated by the company, organization or individual researchers, because some of the conditions mentioned in NIF-ABS format to be signed by the TK holders may not be acceptable or may go against the commercial interest of the TK holders. After CBD Act (1992), India was the first country in the world to implement the Access and Benefit Sharing (ABS) guidelines (Article 8(j)) set forth in CBD Act for equitable sharing of benefit arising out of commercialization of the biodiversity and its associated TK of indigenous community. The Tropical Botanic Garden and Research Institute (TBGRI), Kerala has demonstrated the model that ABS could be possible only when the scientists does not have individualistic view of benefit. The group of scientist lead by P. Pushpangadan searched out small herbs called *Trichopus zylanicus* subsp. *tranvancoricus* Burkill ex Narayanan, locally known as Arogyapacha was discovered from the forest of Western Ghats. The leave of this herb was found to be used by the *Kani* tribe of

Kerala (Western Ghat forest) as anti-fatigue and health rejuvenating agent. Based on these ethnobotanical information, lead scientist of Regional Research Laboratory (RLL), Jammu has discovered various biodynamic compounds notably certain glycolipids and non-steroids compounds with profound adaptogenic and immuno-enhancing properties. The RLL Jammu has filed two patents on the same. In the meantime, the technology of herbal formulation based on TK of TZT was perfected at the laboratory of TBGRI and the same technology was jointly patented in the name of TBGRI and Kani Community. TBGRI transferred the compound purification and drug formulation technology to the Arya Vaidya Pharmaceutical Company (AVPC) for necessary clinical studies to be done on animal model for several time. The Arya Vaid Pharmaceutical Company took a lead role in further clinical evaluation; manufacturing and supply of drug "Jeevani" to the consumers and here Arya Vaid Pharmaceutical Company in legal documents has agreed to pay license fee and 2 per cent royalty to TBGRI from the profit arise out of the commercialization of Jeevani. After getting such profit share from AVPC, the TBGRI resolved to share 1:1 of the license fee and royalty with the Kani tribe (the real TK holders of the TZT). With the help of some local NGOs, TBGRI Scientists and some local government officials, Kani people were encouraged to form a trust consists of adult Kani as its members and the trust was fully owned and managed by the Kani tribe. By the end of year 2005, almost 60 per cent of the Kani families of Kerala were the members of the trust. As per the rules of the TBGRI and Kani Trust, the license fee and royalty received on account of the sale of Jeevani drug would be kept in fixed deposit as corpus fund and only the interest accrued from this amount would be utilized for the benefit and welfare of the members of Kani tribe. Now, the life style of Kani Tribe has improved drastically within 10 years duration. In addition to the license fee and royalty that Kani trust is receiving, a large number of Kani families are getting benefit from the cultivation of Arogyapacha (TZT) and supply of raw materials that is leaves of the plants to the pharmaceutical company for the production of drug Jeevani-the life giver. Many Kani families have been trained by TBGRI scientist for scientific cultivation of the Arogyapacha in and around their community forest land and home Garden (Puspangadan, 1998, 2002). TBGRI model has also given an ethical and moral lesson to the world that scientist or scientific community working on bioresources and associated TK related sector should strictly adhere to CBD's PIC and ABS clause. Individual scientist or sponsoring company must determine to recognize the IPR of traditional knowledge holders (TKH) by involving them as vital components in their research and developmental activities for greater human welfare. But so far, such benefit sharing model and research ethics are yet to be witnessed in NE Region of India. There is an urgent need for scrutiny of existing IPR law at local, regional, national and international level so that interest of the community rights over its biodiversity and TK including phytomedicine related TK could be ensured. Blanket protection of TK holders and community rights through strict IPR law would act as deterrent to further exploitation and pirating of bioresource related TK if policy makers really think of bringing the tribal communities to the mainstream of the globalised world order. State Biodiversity Board (SBB) must frame a comprehensive biodiversity and IPR law as per the requirement of the concerned state while following the guidelines of Indian Biological Diversity Act (2002) and Indian Patent Act (2005). The SBB must also establish IPR cell to strictly implement the SBB-IPR rules to check the infringement. However, to date, only few states of NE India could come up with state biodiversity acts after nearly one and half decades of CBD acts and few late riser states including Arunachal Pradesh are still in preliminary draft stage.

Acknowledgement

The first author is thankful to Ashoka Trust for the Research in Ecology and Environment (ATREE) through CEPF under East Himalayan Small Grant Programme for funding support. Authors also thankful to Department of Science and Technology (DST), New Delhi for funding support through Women Scientist Fellowship Scheme awarded to second author.

References

Albert, L.S. and Kuldip, G., 2006. Traditional use of medicinal plants by the Jaintia tribes in North Cachar Hills district of Assam, northeast India. *Journal of Ethnobiology and Ethno-medicines*, 2: 33 doi: 10.1186/1746-4269-2-33.

Ashok, K.B., 2000. Geophysical Base of Northeast India. *Journal of Assam Science Society*, 41(4): 247–254.

Borang, A., 1996. Studies on certain ethnozoological aspects of *Adi* tribes of Siang district, Arunachal Pradesh, India. *Arunachal Forest News*, 14(4): 1–5.

Chatterjee, S., Saikia, A., Dutta, P., Ghose, D., Panging, G. and Goswami, A.K., 2006. *Background Paper on Biodiversity Significance of North East India, Forest Conservation Programme*, WWF–India, New Delhi.

Cormae, S., 2005. EPO Neem patent revocation revives biopiracy debate. *Nature Biotechnology*, 23(5): 511–512.

Deepak, J.S., 2008. Protection of traditional handicrafts under Indian Intellectual Property Laws. *Journal of Intellectual Property Rights*, 13: 197–207.

Doha, WTO Ministerial Declaration 2001. WT/MIN (01)/DEC/1, 20 November 2001. Ministerial Declaration adopted on 14th November 2001. http://www.wto.org/english/thewto_e/minist_e/min01/_e/mindecl_e.html.

Dutta, B.K. and Dutta, P.K., 2005. Potential of ethnobotanical studies in Northeast India: An overview. *Indian Journal of Traditional Knowledge*, 4(1): 7–14.

FAO, 2004. *International Treaty on Plant Genetic Resources for Food and Agriculture* (FAO of United Nations, Rome, Italy came into force since 29th July 2004.

Hegde, S.N., 2000. Conservation of northeast flora. *Arunachal Forest News*, 18(1 and 2): 10–22.

http://www.sristi.org/honeybee.html for NIF update and explanation note for PIC addressed to the innovators and traditional knowledge holders.

Jeffrey, Collin, 2007. The Patent Act No. 39 of 1970, India Code 2003, The Patent Amendment Act 2005, coming into compliance with TRIPS: A discussion of India's New Patent Laws. *Cardozo Arts and Entertainment Law Journal*, 25(2): 877, 883.

Joshi, P.C. and Joshi, N., 2004. *Biodiversity and Conservation*. APH Publishing Corporation, New Delhi, pp. 382–384.

Karthikeyan, S., 2000. A statistical analysis of flowering plants of India. In: *Flora of India Intro., Vol. 2*, (Ed.) B.D. Sharma. BSI, Kolkata (India), pp. 201–217.

Mao, A.A. and Hyniewta, T.M., 2000. Floristic diversity of Northeast India. *Journal of Assam Science Society*, 41(4): 255–266.

Mashelkar, R.A., 2001. Intellectual property rights and the third world. *Current Science*, 81(8): 955–965.

Myers, N., 1988. Threatened biotas: Hotspots in tropical forest. *The Environmentalist*, 8: 187–208.

Myers, N., Muttermeier, R.A., Muttermeier, C.A., da Fonseca, G.A.B. and Kent, J., 2000. Biodiversity hotspots for conservation priorities. *Nature*, 403, 853–858.

NBA, 2004. The Biological Diversity Act 2002 and Biological Diversity Rules 2004. National Biodiversity Authority, Chennai (India), p. 56–57.

Nima, N.D., Tag, H., Mandal, M., Kalita, P. and Das, A.K., 2009. An ethnobotanical study of traditional anti-inflammatory plants used by the Lohit Community of Arunachal Pradesh, India. *Journal of Ethnopharmacology*, 125: 235–245.

Patwardhan, B., 2007. *Drug Discovery and Development: Traditional Medicine and Ethnopharmacology*, New India Publishing Agency, Pitam Pura, New Delhi, India, pp. 7–322.

Puspangadan, P. and Nayar, K.N., 2005. Access to biological resources and associated traditional knowledge and benefit sharing. An extended abstract in Ministry of Environment and Forest, *The group of likeminded megadiverse country: Realizing the potential*, MoEF, Govt of India, pp. 10–13.

Puspangadan, P., 1998. Ethnobiology in India: A status report. All India Coordinated Research Project on Ethnobiology, Ministry of Environment and Forest, Govt. of India, New Delhi, p. 68.

Puspangadan, P., 2002. Biodiversity and emerging benefit sharing arrangement: Challenges and opportunities for India. *Proceeding of Indian National Science Academy (P–INSA)* B–68, pp. 297–314.

Sahai, Suman, 2003. India's plants varieties protection and farmers right acts 2001. *Current Science*, 84(3): 407–412.

Shiva, V., Bhar, R.H. and Jafri, A.H., 2002. Corporate hijack of biodiversity. How WTO–TRIPS Rules promote corporate hijack of people's biodiversity and knowledge. Navadanya Publication, New Delhi.

Sumit, C., Gopal, S., Suman, M. and Suresh, C.P., 2008. Farmer's right in conserving plant biodiversity with special reference to North East India. *Journal of Intellectual Property Rights*, 13: 225–233.

Tag, H. and Das, A.K., 2004. Ethnobotanical notes on Hill Miri (Nyishi) tribes of Arunachal Pradesh. *Indian Journal of Traditional Knowledge*, 3(1): 80–85.

Tag, H., Murtem, G., Das, A.K. and Singh, R.K., 2008. Diversity and distribution of ethno-medicinal plants used by the Adi Tribe in East Siang District of Arunachal Pradesh, India. *Pleione*, 2: 123–136.

Tag, H., Nima, N.D., Das, A.K., Kalita, P. and Mandal, S.C., 2009. Evaluation of anti-inflammatory potential of *Chloranthus erectus* (Buch-Ham.) Verd leaf extract in rat. *Journal of Ethnopharmacology*, 126: 371–374.

The WIPO Intergovernmental Committee on Intellectual Property, Genetic Resources, Traditional Knowledge and Folklore (IGC) was established by the WIPO General Assembly in October 2000 (http://www.wipo.int/documents/en/documents/govbody/wo_gb_ga/pdf/ga26_6.pdf.

TRIPS, Annex IC of 1994 Marrakesh Agreement establishing World Trade Organisation; came into force on 1st January 1995.

UNEP, Convention on Biological Diversity, Rio de Janeiro, 1992, came into force w.e.f. 29th December 1993. http://www.biodiv.org/doc/legal/cbd.

Chapter 4

Isolation and Characterization of Active Principles from *Artemisia caruifolia* Roxb.

Padma Acharyya and N.C. Baruah

ABSTRACT

The over ground parts of Artemisia caruifolia of compositae family were chemically examined which led to the isolation of one new and five known coumarins along with previously isolated dihydro cinnamic acid. These are named as compounds A-G for the purpose of discussion.

Keywords: Artemisia, Sesquiterpene lactone, coumarins.

Introduction

The genus *Artemisia* is one of the largest and most widely distributed plant in the temperature, sub-temperature, and tropical regions of the world. It has as many as 300 species. Some of the plants grow wild as weed and others are cultivated for economic use. Many of them are medicinal and are the source of the valuable anthelmintic drugs, *e.g.* santonin (0.5–3 per cent). Several species yield essential oils and few are reported to be useful as fodder.[1]

Sesquiterpene lactones are characteristic constituents of the plants of compositae family[2] although they are very rarely found in other plant families also. Sesquiterpene lactones contains $\alpha:\beta$ unsaturated γ-lactone moiety as a major structural feature. Recent studies have revealed that this moiety is associated with carcinogenic, antitumer and other biological activities.Chemical investigation of *Artemisia caruifolia* was carried out in search of Sesquiterpene lactones and isolated six coumarins (A-F).

Results and Discussion

The least polar compound A was obtained as a white crystal, mp 184°-187° C and was identified as daphnetin methylene ether 1 by direct comparison with an authentic sample[3] (T.L.C., ^{1}HNMR, MS, IR).

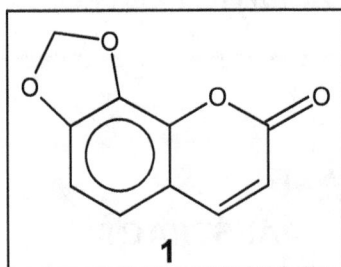

1

The new compound B was obtained as a gum which displayed IR band at 1720 cm[-1] suggested it to be a coumarin. Its ^{1}HNMR spectrum displayed a pair of doublets at δ 7.56 ppm with J=9.4 Hz and at δ 6.14 ppm with J=9.5 Hz integrating to one proton each and a singlet at δ 7.06 ppm integrating to one proton. This pattern of NMR signals was typical of 6, 7, 8-trisubstitutd coumarins. The ^{1}HNMR spectrum further revealed the presence of a doublet at δ 4.62 ppm with J=7 Hz integrating to one proton at C-1$^{/}$, a triplet at δ 5.41 ppm with J=6.8 Hz integrating to two protons at C-2$^{/}$ and a broad signal at δ 1.78 ppm integrating to six protons thus indicating the presence of a prenyloxy moiety in the molecule. A singlet at δ 3.41 ppm integrating to six protons indicated the presence of two methoxyl groups at C-6 and C-8. The mass spectrum of B gave a molecular ion peak at m/z 313 (M[+]+Na) indicating a molecular weight of 290. Based on all these data, the structure of the compound has been characterized as 2.

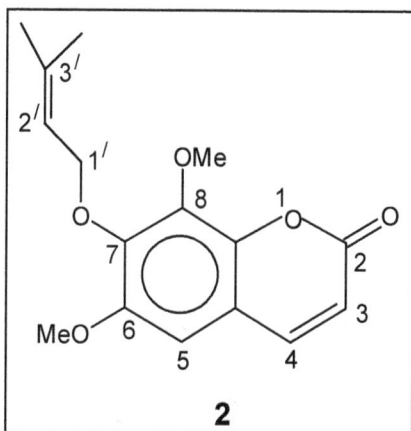

2

The next polar compound C mp 93°C was identified as daphnetin-7-methyl-8-(3,3-dimethyl allyl) ether 3 by direct comparison (T.L.C., NMR, IR and MS) with its authentic sample.[4]

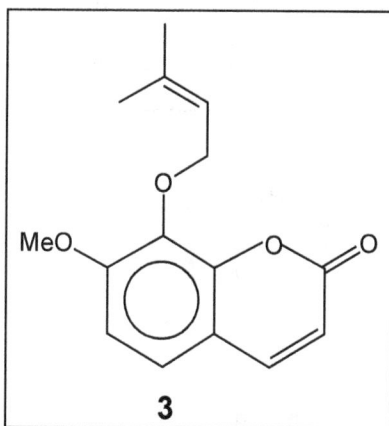

3

Compound D was crystallized from ethylacetate, mp 176°C and was identified as daphnetin-7-methyl ether 4 by comparing the T.L.C. and ^{1}H NMR spectrum with the authentic sample[5] which were indistinguishable.

Compound E, mp 116°-118°C was identified as daphnetin dimethyl ether 5 by direct comparison (T.L.C. and NMR) with an authentic sample.[5]

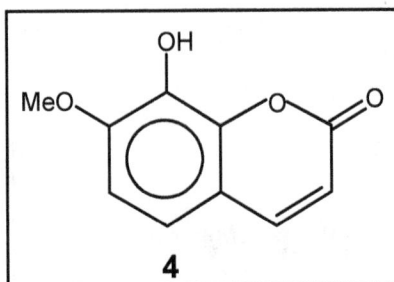

4

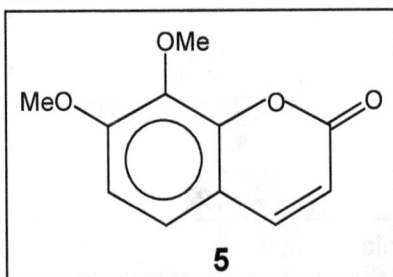

5

The compound F was crystallized from chloroform – hexane, mp 133°C was identified as 3,4-dimethoxy-2-hydroxy cinnamic acid 6 by direct comparison (T.L.C. and NMR) with an authentic sample.[4]

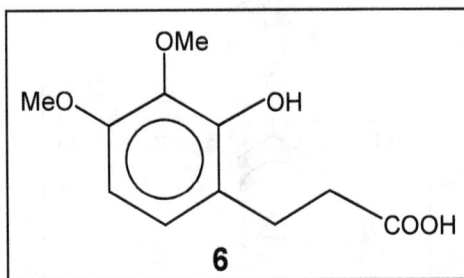

6

The most polar compound G of this series was a crystalline solid, mp 140°C (lit.[6] mp 148°C), isolated for the first time from this plant. The mass spectrum of this compound gave a molecular ion peak at m/z 244.9 ($M^+ + Na$) indicating a molecular weight of 222.8. It was characterized as 7-hydroxyl-6, 8-dimethoxy coumarin 7 from its spectral data (NMR, IR, MS) which are in good agreement with those reported in the literature.[7]

OMe

HO

O

O

MeO

7

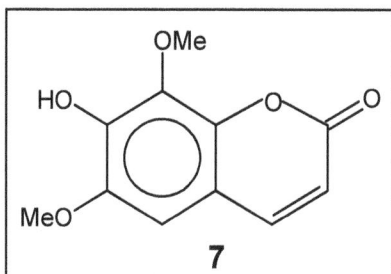

Experimental

Above ground parts of *Artemisia caruifolia* (2 kg) collected from the Janjimukh area of Sibsagar district, Assam, India, were extracted with chloroform in a soxhlet apparatus until the extract was colourles. After evaporation of chloroform under reduced pressure, the residue (52 g) was dissolved in 300 mL methanol containing 30 mL water and left overnight at room temperature. It was then filtered and the filtrate was washed thoroughly with petroleum ether (bp 60°-80°C, 6 × 300 mL). The methanol portion was concentrated at reduced pressure and then extracted with chloroform (5 x 200 mL). Evaporation of the washed and dried extract of both petroleum ether and chloroform furnished 10 g and 20 g of gummy residue respectively. The combined residue was chromatographed over 600 g of silica gel-G in a column, packed in hexane and then different fractions (200 mL each) were collected for futher investigation.

Fractions 45-51 showed a single spot on T.L.C., were combined and purified by preparative T.L.C. (EtOAc: Hexane, 1:8) to yield 35 mg of 1 as a crystalline solid.

mp : 187°C.

^{1}H NMR (300 MHz, CDCl$_3$) : δ 6.03 (s, -O-CH$_2$-O-), 6.06 (d, J= 9.5 Hz, 1H, H-3), 6.9 (d, J= 8.3 Hz, 1H, H-6), 7.03 (d, J=8.5 Hz, H, H-5), 7.74 (d, J=9.3 Hz, 1H, H-4).

IR (CHCl$_3$) : 1729, 1633, 1580, 1275, 1185, 1075, 910 cm^{-1}.

MS (m/z) : 190 (M$^+$).

Direct comparison with an authentic sample[3] by T.L.C. and NMR identified the compound as 1.

Fractions 84-92 showed one minor and one major spot on T.L.C. (iodine exposure) were combined and purified by preparative T.L.C. (EtOAc: Hexane, 1:5). The faster moving band yielded 55 mg of 2 as gum.

¹H NMR (300 MHz, CDCl₃) : δ 7.56 (d, J= 9.4Hz, 1H, H-4), 7.06 (s, 1H, H-5), 6.14 (d, J=9.5 Hz, 1H, H-3), 5.41 (t, J=6.8 Hz, 1H, H-2′), 4.62 (d, J=7 Hz, 2H, H-1′), 3.41 (s, 6H,-OCH₃), 1.78 (br, vinyl methyls).

IR (CHCl₃) : 3390, 1720, 1576, 1509, 1415, 1300, 1122, 1040, 980, 850 and760 cm⁻¹

MS (m/z) : 313(M⁺ + Na), 290 (M⁺)

The slower moving band yielded 110 mg daphnetin-7- methyl-8-(3,3- dimethyl allyl) ether 3.

mp : 93°C.

¹H NMR (300 MHz, CDCl₃) : δ 7.55 (d, J= 9.5 Hz, 1H, H-4), 7.05 (d, J=8.5 Hz, 1H, H-5), 6.76 (d, J= 8.5 Hz, 1H, H-6), 6.12 (d, J= 9.5 Hz, 1H,H-3), 5.39 (t, J= 6.7 Hz, 1H, H-2/), 4.60 (d, J=7Hz, 2H, H-1), 3.93 (s, 3H, -OCH₃) 1.75 br (vinyl methyls).

IR (CHCl₃) : 1730, 1644, 1603, 1275, 1185, 1120, 1070 and 825 cm⁻¹.

MS (m/z) : 260 (M⁺).

These values tallied exactly with those of an authentic sample[4] 3 to confirm its identity.

Fractions 97-115, which exhibited one major spot on T.L.C. were combined and purified by preparative T.L.C. (EtOAc:Hexane, 1:3) to yield 30 mg of daphnetin-7-methyl ether 4 as colorless crystals.

mp : 176°C (lit.[5] mp 178° C).

¹H NMR (300 MHz, CDCl₃) : δ 4.10 (s, 3H,-OCH₃), 6.13 (d, J= 9.5 Hz, 1H, H-3), 6.82 (d, J= 8.5 Hz, 1H, H-6), 7.01 (d, J= 8.3Hz, 1H, H-5), 7.49 (d, J=9.3 Hz, 1H, H-4).

IR (CHCl$_3$) : 1727, 1610, 1573, 1453, 1405, 1166, 1080, 1020 and 965 cm^{-1}.

MS (m/z) : 192 (M$^+$).

NMR spectrum of this compound was indistinguishable from those of authentic sample 4.[5]

Fractions 125-129 were combined to give 250mg of crude material which was purified by preparative T.L.C. (EtOAc: Hexane, 1:1) to furnish 110 mg of daphnetin dimethyl ether 5 as colourless needles.

mp : 116 -118°C (lit.6 mp119 - 120° C).

^{1}H NMR (300 MHz, CDCl$_3$) : δ 3.87 (s, 3H, -OCH$_3$), 3.83 (s, 3H, - OCH$_3$), 6.16 (d, J=9.3 Hz,1H,H-3), 6.81 (d, J= 8.4 Hz, 1H, H-6), 7.29 (d, J= 8.5 Hz, 1H, H-5), 7.58 (d, J= 9.2 Hz, 1H, H-4).

IR (CHCl$_3$) : 1725, 1603, 1562, 1499, 1290, 1125, 1055 and 825 cm^{-1}.

MS (m/z) : 206 (M$^+$).

Fractions 142-150 exhibited one major spot on T.L.C. were combined and recrystallised from chloroform- hexane afforded 77 mg of 3,4-dimethoxy-2-hydroxyl cinnamic acid 6, mp 133°C identified by direct comparison (T.L.C., NMR and MS) with an authentic sample.[4]

Fractions 163-167 showed one minor spot on T.L.C. and on purification by preparative T.L.C. (EtOAc : Hexane, 3:1) gave 40 mg of 7 as crystalline solid.

mp : 140° C (lit[6] mp148° C).

^{1}H NMR (300 MHz, CDCl$_3$) : δ 4.06 (s, 3H, -OCH$_3$), 3.93 (s, 3H, - OCH$_3$), 6.28 (d, J= 9.32 Hz, 1H, H-3), 7.61 (d, J= 9.42 Hz, 1H, H- 4), 6.35 (s, 1H, H-5).

IR (CHCl$_3$) : 3396, 3012, 2945, 1709, 1574.5, 1500, 1418, 1310, 1157, 1122, 1037, 978, 847 and 756 cm^{-1}.

MS (m/z) : 244.9 (M$^+$ + Na), 222.8 (M$^+$)

Fractions 168 – 230 did not show any promising spots on T.L.C.

References

Barua, N.C., Sharma, Ram P., Madhusudanan, K.P., Thyagarajan, G. and Herz, W., 1980. *Phytochemistry*, 19: 2217.

CSIR, 1948. *The Wealth of India*. CSIR Publications, New Delhi, p. 120.

Dean, F.M. (ed.), 1963. *Naturally Occurring Oxygen Ring Compounds*. Butterworth and Co. (Publishers) Limited, London.

Herz, W., Bhat, S.V. and Santhanam, P.S., 1970. *phytochemistry*. 9: 891.

Kelsey, R.G. and Shafizadeh, 1979. *Phytochemistry*, 18: 1591.

Pavanasasivam, G., Sultanbawa, M. and Uvais, S., 19753 *J. Chem. Soc. Perkin–I*, 19: 612.

Sham'yanov, I.D., Mallabaev, A. and Sidyakin, G.P., 1974. *Khim. Prir. Soedin.*, p. 784.

Chapter 5

The Role of Medicinal Plants in Traditional Medicine and Current Therapy

Manika Das

Background

Biodiversity of natural resources harbor many secrets for health care and are evident from the past history of human civilization. The ancient text of *Ayurveda* reports more than 2000 plant species for their therapeutic potentials. Traditional medicines of *Ayurveda, Siddha, Unani* and homeopathy comprise a wide range of therapeutic approaches. Medicinal plants are used as a source of therapeutic agents as they contain wide variety of bioactive compounds *i.e.* digoxin, digitoxin, morphin etc. About 6 per cent of total known higher and lower plant species of this planet have been screened for biologic activity, and reported 15 per cent have been evaluated phytochemically (Verpoorte, 2000).

Medicinal plants can be promising sources of natural products with potential anticancer, antimicrobial and antioxidative activity.

* Corresponding author: E-mail: mkdas116@gmail.com

Results from various researches on bioactive components of different plant sources will guide the selection of plant species for further pharmacological and phyto-chemical investigation. According to WHO almost 65 per cent of the world's population have in incorporated into their modality of health care. There is revival of interest in drug discovery from medicinal plants for the maintenance of health in all parts of the world.

Plants have formed a basis for traditional medicine systems that have been used thousands of years in countries *i.e.* China, India, Thailand etc. The use of plants in traditional medicine systems of many other cultures has been extensively documented (Schultes and Raffauf 1990; Arvigo and Balick 1993, Gupta 1995). Although the study of medicinal plants of traditional medicine was of no interest to the government, academic and health institutions in U.S. and Europe, the fact is that scientific interest to evaluate the healing properties of such natural resources grew rapidly among some groups of intellectuals and scientists in Asia, Africa and Latin America. The strategy of promoting industrial and commercial development of those resources was predominantly in China, India and Japan, where the possibility soon raised that the same plants used empirically for centuries in their local cultures they made drugs worldwide, whose effectiveness was scientifically proven and promoting the development of new types of 'Plant medicines' or also known as 'Phyto-medicines'.

Diversity of secondary plant metabolites that results from plant evolution may be equal or superior to that found in synthetic combinatorial chemical libraries, as the later involve adequate chemical knowledge, processing and the risk of production of toxic side effects.

Traditional Medicine Inspired Approaches to Drug Discovery from Medicinal Plants

Various reports on drug discovery document the value of using traditional medical information to initiate drug discovery efforts. The WHO-TRM (Traditional Medicine Programme) centers throughout the world assist in identifying all plant-derived pure compounds used as drugs in their respective countries. Study on scientific literature reporting isolation of plant derived bioactive compounds has been done to determine whether the chemical efforts were stimulated by traditional medical or ethno-medical claims and to correlate current uses for the compounds with such claims (Farnsworth1985). A total of 122 compounds were identified; 80 per cent of these compounds were used for the ethno-medical purposes (Table 5.1). Further, it was discovered that these compounds were derived from 94 species of plants (Farnsworth *et al.,* 1985) and a conservative estimate of the number of flowering plants occurring on the planet is about 250,000. So, there should be an abundance of drugs remaining to be discovered in these plants. The question is what can be best approach to discover plants containing potential drugs?

In 1985, the WHO Special Programme of Research and Training in Human Reproduction embarked on a programme called "The Task Force on Plants for Fertility regulation" (Spieler 1981). The charge was to select plants on the basis of ethno-medical claims related to human reproduction *e.g.* abortifacient, contraceptive etc. Safety with long-term use was presumed. The ultimate goal was to discover orally active pure substances that were non-esterogenic, non-steroidal and non-toxic anti-plantation agents. Work was carried out initially in designated centers in U.S., England, South Korea, Brazil, India and Hongkong, with additional centers later established in People's Republic of China and Thailand. The work involved searching of all available literature for plants and natural compounds having any of these biologic effects and storing these informations for eventual analysis (Farnsworth 1981). Approximately 4,000 plant species were identified, 300 species were scheduled for collection and testing. Several active compounds were identified, the most promising was an indole alkaloid named Yuehchukene (YCK) (Kong *et al.,* 1985) from the plant *Murraya paniculata*(L), used in China to regulate fertility. Unfortunately, YCK showed a low level of esterogenicity and WHO program was terminated shortly thereafter.

In 1985, an approach based on ethno-medical information, to experimentally pursue plants as a source of drugs was proposed (Fabricant 2001). The approach was designed primarily for implementation by developing countries, where shortage of finance often prevents sophisticated types of research from being conducted.

Table 5.1: Drugs Derived from Plants, with their Ethno-medical Correlations and Sources*

Drug	Action or Clinical Use	Plant Source
Acetyldigoxin	Cardiotonic	*Digitalis lantana* Ehrh.
Adoniside	Cardiotonic	*Adonis vernalis* L.
Aescin	Anti-inflammatory	*Aesculus hippocastanum* L.
Aesculetin	Antidysentery	*Fraxinus rhynchophylla* Hance
Agrimophol	Anthelmintic	*Agrimonia eupatoria* L.
Ajmalicine	Circulatory disorders	*Rauvolfia serpentine* (L) Benth ex. Kurz
Allyl isothiocyanate	Rubefacient	*Brassica nigra* (L.) Koch
Andrographolide	Bacillary dysentery	*Andrographis paniculata* Nees
Anisodamine	Anticholineric	*Anisodus tanguticus* (Maxim.) Pascher
Anisodine	Anticholineric	*Anisodus tanguticus* (Maxim.) Pascher
Arecoline	Anthelmintic	*Arecha catechu* L.
Asiaticoside	Vulnerary	*Centella asiatica* (L.)Urban
Atropine	Anticholinergic	*Atropa belladonna* L.
Berberine	Bacillary dysentery	*Berberis vulgaris* L.
Bergenin	Antitussive	*Ardisia japonica* Bl.
Bromelain	Anti-inflammatory; proteolytic	*Ananas comosus* (L.) Merrill
Caffeine	CNS stimulant	*Camellia sinensis* (L.) Kuntze
(+)Catechin	Haemostatic	Potentilla fragaroides L.
Chymopapain	Proteolytic; mucolytic	Carica papaya L.
Cocaine	Local anaesthetic	*Erythroxylum coca* Lamk.
Codeine	Analgesic; antitussive	*Papaver somniferum* L.
Colchicine	Antitumor agent	*Colchicum autumnale* L.
Convallotoxin	Cardiotonic	*Convallaria majalis* L.
Curcumin	Choleretic	*Curcuma longa* L.
Cynarin	Choleretic	*Cynara scolymus* L.
Danthron	Laxative	*Cassia* spp.
Deserpidine	Antihypertensive; tranquilizer	*Rauvolfia canescens* L.

Contd...

Table 5.1–Contd...

Drug	Action or Clinical Use	Plant Source
Deslanoside	Cardiotonic	*Digitalis lantana* Ehrh.
Digitalin	Cardiotonic	*Digitalis purpurea* L.
Digitoxin	Cardiotonic	*Digitalis purpurea* L.
Digoxin	Cardiotonic	*Digitalis lantana* Ehrh.
Emetine	Amoebicide	*Cephaelis ipecacuanha* (Brotero) A.
Ephedrine	Sympathomimetic	*Ephedra sinica* Stapf.
Etoposide	Antitumor agent	*Podophyllum peltatum* L.
Gitalin	Cardiotonic	*Digitalis purpurea* L.
Glaucaroubin	Amoebicide	*Simarouba glauca* DC.
Glycyrrhizin	Sweetner	*Glycyrrhiza glabra* L
Hemsleyadin	Bacillary dysentery	*Helmsleya amabilis* Diels
Hydrastine	Hemostatic; astringent	*Hydrastis canadensis* L.
Hyoscamine	Anticholinergic	*Hyoscamus niger* L.
Kainic acid	Ascaricide	*Digenea simplex* (Wulf.) Agardh
Kawain	Tranquilizer	*Piper methysicum* Forst. f.
Khellin	Bronchodilator	*Ammi visnaga* (L.) Lamk.
Lanatosides A,B,C	Cardiotonic	*Digitalis lantana* Ehrh.
Lobeline	Respiratory stimulant	*Lobelia inflate* L.
Monocrotaline	Antitumor agent	*Crotolaria sessiliflora* L.
Morphine	Analgesic	*Papaver somniferum* L.
Neoandrographolide	Bacillary dysentery	*Andrographis paniculata* Nees
Noscapine	Antitussive	*Papaver somniferum* L.
Ouabain	Cardiotonic	*Strophanthus gratus* Baill.
Papain	Proteolytic; mucolytic	*Carica papaya* L.
Phyllodulcin	Sweetner	*Hydrangea macrophylla* (Thunb.) DC
Physostigmine	Cholinesterase inhibitor	*Physostigma venenosum* Balf.
Picrotoxin	Analeptic	*Anamirta cocculus* (L.) W&A
Pilocarpine	Parasympathomimetic	*Pilocarpus jaborandi* Holmes
Podophyllotoxin	Conylomata acuminata	*Podophyllum peltatum* L.
Pseudoephedrine	Sympathomimetic	*Ephedra sinica* Stapf.

Contd...

Table 5.1–Contd...

Drug	Action or Clinical Use	Plant Source
Quinine	Antimalarial	*Cinchona ledgeriana* Moens ex. Trimen
Quisqualic acid	Anthelmintic	*Quisqualis indica* L.
Rescinnamine	Antihypertensive; tranquilizer	*Rauvolfia serpentine* (L.)Benth ex. Kurz
Reserpine	Antihypertensive; tranquilizer	*Rauvolfia serpentine* (L.)Benth ex. Kurz
Rhomitoxin	Antihypertensive	*Rhodendron molle* G. Don
Salicin	Analgesic	*Salix alba* L.
Santonin	Ascaricide	*Artemisia maritime* L.
Scillarin A	Cardiotonic	*Urginea maritima*
Scopolamine	Sedative	*Datura metel* L.
Teniposide	Antitumor agent	*Podophyllum peltatum* L.
Theophylline	Diuretic; bronchodilator	*Camellia sinensis* (L.) Kuntze
Trichosanthin	Abortifacient	*Thymus vulgaris*
Tubocurarine	Skeletal muscle relaxant	*Chondodendron tomentosum* R.& P.
Vincamine	Cerebral stimulant	*Vinca minor* L.
Xanthotoxin	Leukoderma; vitiligo	*Ammi majus* L.
Yuanhuadine	Abortifacient	*Daphne genkwa* Seib.& Zucc.

* Data adapted from Farnsworth *et al.* (1985).

Approaches to Drug Discovery Using Higher Plants

Several reviews pertaining to approaches for selecting plants as candidates for drug discovery programs have been published (Verpoorte 2000; Phillipson 1989; Kinghorn 1994; Vlietinck 1991; Harvey 2000; Suffness 1982).

Phytochemical Screening Approaches

Random selection of plants followed by chemical screening for the presence of alkaloids, triterpenes, flavonoids, isothiocyanates etc. can be done. These have been used in past and are currently pursued mainly in the developing countries.

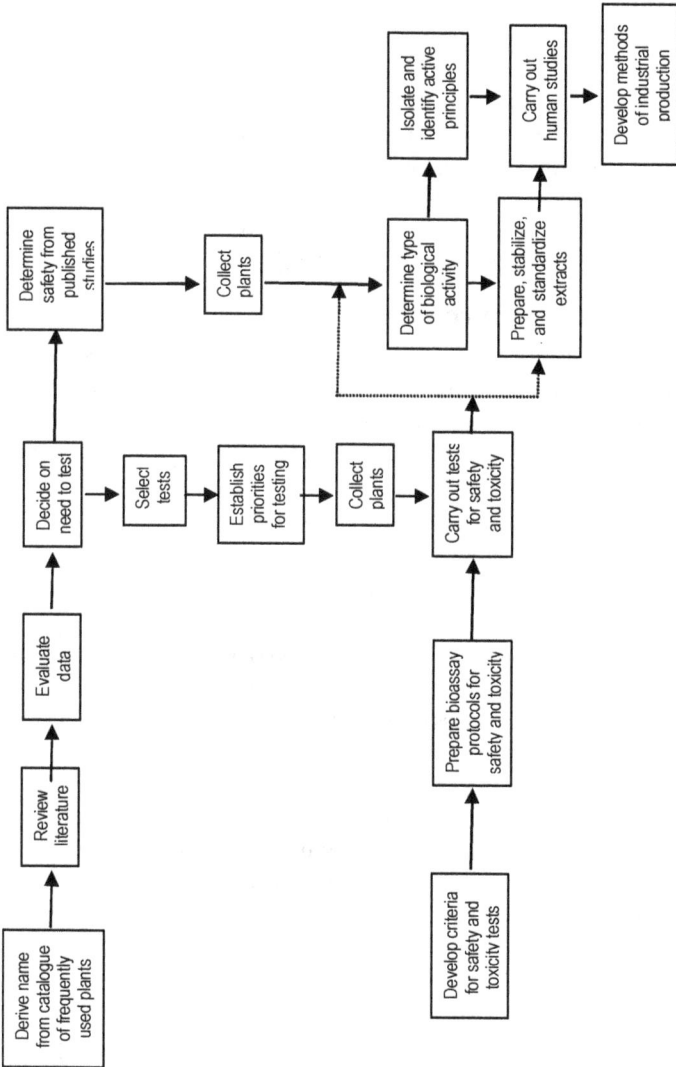

Figure 5.1: Flow Chart of Sequence for the Study of Plants Used in Traditional Medicine (Adapted from Farnsworth *et al.*, 2001)

Biological Assay Approaches

Random selection followed by one or more biologic assays were done in different programs, were sponsored by the National Cancer Institute (NCI) (Suffness 1982) in the United States and the Central Drug Research Institute (CDRI) of India (Bhakuni 1969). The CDRI evaluated approximately 2,000 plant species for several biologic activities including antibacterial, anti-diabetic, anti-fertility, antifungal, antitumor, diuretic, and others (Dhar 1968). Follow up of biologic activity reports of traditional medicinal plants is important, as the plant extract posses interesting biologic activity, and their active principles need to be studied.

Ethno-medical Approaches

Ayurveda, Unani, Kampo and traditional Chinese medicines have flourished as systems of medicine in use for thousands of years. Their individual arrangements emphasize education based on an established, frequently revised body of written knowledge and theory. These systems are still in place today because of their organizational strengths, and they focus primarily on multi-component mixtures (Bannerman 1983).

Herbalism, folklore and *shamanism* centre on an apprenticeship system of information passed to the next generation through a *shaman*, traditional healer, or herbalist. The plants that are used are often kept secret by the practitioner, so little information about them is recorded; thus, there is less dependence on scientific evidence as in systems of traditional medicine that can be subjected to scrutiny. The *shaman* or herbalist combines the roles of pharmacist and medical doctors with the cultural or spiritual religious beliefs of a region or people which are regarded as magic or mysticism. This approach is widely practiced in Africa and South America (Rastogi 1982). Ethno-medical information can be acquired from various sources such as books on medical botany (Lewis *et al.,* 1977) and herbals (Cruz *et al.,* 1940); review articles (Siddiqui *et al.,* 2000); field work (Soejarto 1989); and computer databases (69-71).

Value of Ethno-medicinal Plants in Current Therapy

Astruso (Artuso1997) has outlined the entire process of using plants for drug discovery: formulating an appropriate strategy, obtaining biologic extracts, screening those extracts, isolating active compounds, conducting preclinical tests and chemical modification, submitting an Investigational New Drug Application, performing clinical trials, submitting a New Drug Application, and beginning commercial production. He estimates the entire process would take 10-20 years or more. Some examples of drugs from plants that served as models for the next generation of drugs are exemplified in Table 5.2.

Table 5.2: Plant-Derived Drugs and their Sources Not Developed on the Basis of Ethno-medical Information

Drug	Plant Source
Allantoin	Several plants
Anabasine	*Anabasis aphylla* L.
Benzyl benzoate	Several plants
Borneol	Several plants
Camphor	*Cinnamonum camphora* (L.) J.S. Presl
Camptothecin	*Camptotheca acuminata* Decne.
Cissampeline	*Cissampelos pareira* L.
Colchicaine amide	*Colchicum autumnale* L.
Demecolcine	*Colchicum autumnale* L.
L-Dopa	*Mucuna deeringiana* (Bort) Merr.
Galanthamine	*Lycoris squamigera* Maxim.
Glaucine	*Glaucium flavum* Crantz
Glaziovine	*Ocotea glazovii* Mez
Hesperidin	*Citrus spp.*
Huperzine A	*Huperzia serrata* (Thunb. ex Murray) Trevis.
Menthol	*Mentha spp.*
Methyl salicylate	*Gaultheria procumbens* L.
Nicotine	*Nicotiana tabacum* L.
Nordihydroguaiaretic acid	*Larrea divaricata* Cav.
Pachycarpine	*Sophora pachycarpa* Schrenk ex C.A. Meyer
Palmatine	*Coptis japonica* Makino
Papaverine	*Papaver somniferum* L.
Pinitol	Several plants
Quinidine	*Cinchona ledgeriana* Moens ex. Trimen

Contd...

Table 5.2–Contd...

Drug	Plant Source
Rutin	*Citrus* spp.
Sanguinarine	*Sanguinaria candensis* L.
Sparteine	*Cytisus scoparius* (L.) Link
Taxol	*Taxus brevifolia* Nutt.
Tetrahydrocannabinol	*Cannabis sativa* L.
Tetrandrine	*Stephania tetrandra* S.Moore
Thymol	*Thymus vulgaris* L.
Vasicine	*Adhatoda vasica* Nees
Vinblastine	*Catharanthus roseus* (L.) G. Don
Vincristine	*Catharanthus roseus* (L.) G. Don

Data adapted from Farnsworth *et al.* (1985).

The challenges which are faced in traditional medicine are: inadequate knowledge for screening, toxic effect, adequate dose requirement etc. It is time consuming to collect specific plants having an ethno-medical history. The adjuvant of high throughput, mechanism based in vitro bioassays coupled with potential plants derived from ethno-pharmacological research has resulted in the discovery of new pharmaceuticals such as prostratin-a drug for treatment of HIV, as well as variety of antioxidants, antibacterial, antifungal, anti-inflammatory compounds.

The information derived from traditional medicine can be used to provide resolutions in the search for new drugs for current therapy like- as a general indicator of non specific bioactivity suitable for a panel of broad screens:

☆ As an indicator of specific bioactivity suitable for high resolution bioassays,

☆ As an indicator of pharmacological activity for which mechanism based bioassays have yet to be developed.

There are merits and demerits of using plants as the starting material in any drug development program. If one elects to use information based on the ethno-medicine, one can rationalize that any isolated active compounds from the plants are likely to be safer than the active compounds from plants with no history of human use. It is universally believed that plants provide an unlimited source of novel and complex compounds, from which many useful drugs have been discovered by follow up of traditional medicines (Table 5.1).

There is an urgent need for the practitioners of the traditional medicine and current therapy to work together to optimize the risk benefit profile of these medicines.

Several challenges emerge while choosing particular plants, firstly as plants are biologic system and have inherent potential variability in their chemistry and biologic activity. Reports show that 25 per cent of all selected plants showing promising activity fail to have the activity confirmed in biologic assay systems (Farnsworth 1966 and 1977; Farnsworth *et al.*, 1962 and 1966).

Conclusion

The use of ethno-medicinal knowledge in discovery of drugs has led to great developments in healthcare. Rapid industrialization of the planet is resulting into loss of ethnic cultures and customs. An abundance of ethno-medical information on plant uses can be found in the scientific literature, but has not yet been compiled into a usable form.

The medicinal value of phyto-medicine need to be confirmed by scientific studies that contain pre-clinical information and clinical data supporting that they are effective, non-toxic or have adverse effects. Rationally designed, carefully standardized, synergistic traditional herbal formulations and botanical drug products with robust scientifically evidence can be alternatives. However, we still have a long way to go. There is an urgent need for the practitioners of current therapy and traditional medical system to work together to optimize the risk-benefit profile of these medicines.

References

Artuso, A., 1997. *Drugs of Natural Origin: Economic and Policy Aspects of Discovery, Development, and Marketing.* Pharmaceutical Products Press, New York.

Bannerman, R.H.O., Burton, J. and Ch'en, W.-C., 1983. *Traditional Medicine and Health Care Coverage: A Reader for Health Administrators and Practitioners.* World Health Organization, Geneva.

Bhakuni, D.S., Dhar, M.L., Dhar, M.M., Dhawan, B.N. and Mehrotra, B.N., 1969. Screening of Indian plants for biological activity. II. *Indian J. Exp. Biol.*, 7: 250–262.

Clark, A.M., 1996. Natural products as a resource for new drugs. *Pharm. Res.,* 13: 1133–1144.

Cruz, M., Badiano, J. and Trueblood, E.W.E., 1940. The Badianus Manuscript, Codex Barberini, Latin 241, Vatican Library: An Aztec Herbal of 1552. Baltimore, MD: Johns Hopkins Press.

Dhar, M.L., Dhar, M.M., Dhawan, B.N., Mehrotra, B.N. and Ray, C., 1968. Screening of Indian plants for biological activity. I. *Indian J. Exp. Biol.*, 6: 232–247.

Fabricant, D.S. and Fransworth, N.R., 2001. The value of plants used in traditional medicine for drug discovery. *Environmental Health Perspectives*, 109(1): 69–75.

Farnsworth, N.R., Akerele, O., Bingel, A.S., Soejarto, D.D. and Guo, Z., 1985. Medicinal plants in therapy. *Bulletin WHO,* 63: 965–981.

Farnsworth, N.R. and Bingel, A.S., 1977. Problems and prospects of discovering new drugs from higher plants by pharmacological screening. In: *New Natural Products and Plant Drugs with Pharmacological, Biological or Therapeutical Activity,* (Eds.) H. Wagner, P. Wolff. Springer, Berlin, p. 1–22.

Farnsworth, N.R., Henry, L.K., Svoboda, G.H., Blomster, R.N., Yates, M.J. and Euler, K.L., 1966. Biological and phytochemical evaluation of plants. In: *Biological Test Procedures and Results from 200 Accessions. Lloydia,* 29: 101–122.

Farnsworth, N.R., Pilewski, N.A. and Draus, F.J., 1962. Studies on falsepositive alkaloid reactions with Dragendorff's reagent. *Lloydia*, 25: 296–310.

Farnsworth, N.R., 1966. Biological and phytochemical screening of plants. *J. Pharm. Sci.*, 55: 225–276.

Harvey, A., 2000. Strategies for discovering drugs from previously unexplored natural products. *Drug Discovery Today,* 5: 294–300.

Itharat, A. and Ooraikul, B., 2007. Research on Thai medicinal plants for cancer treatment. In: *Advances in Medicinal Plant Research*, p. 287–317.

Kinghorn, A.D., 1994. The discovery of drugs from higher plants. *Biotechnology,* 26: 81–108.

Lewis, W.H. and Elvin-Lewis, M.P.F., 1977. *Medical Botany: Plants Affecting Man's Health.* Wiley, New York.

Mukherjee, P.K. and Wahile, A., 2006. Integrated approaches towards drug development from *Ayurveda* and other Indian systems of medicines. *J. Ethnopharmacology,* 103(1): 25–35.

Phillipson, J.D. and Anderson, L.A., 1989. Ethnopharmacology and Western medicine. *J. Ethnopharmacol.,* 25: 61–72.

Rastogi, R.P. and Dhawan, B.N., 1982. Research on medicinal plants at the Central Drug Research Institute, Lucknow (India). *Indian J. Med. Res.,* 76(Suppl): 27–45.

Siddiqui, T.O., Javed, K. and Alam, M.M., 2000. Folk-medicinal claims of western Uttar Pradesh, India. *Hamdard Med.* and 43: 59–60.

Soejarto, D.D., 1993. Biodiversity prospecting and benefit-sharing: Perspectives from the field. *J. Ethnopharmacol.,* 51: 1–15.

Soejarto, D.D., 1993. Logistics and politics in plant development. In: *Human Medicinal Agents from Plants.* ACS Symposium Series, No. 534, (Eds.) A.D. Kinghorn and M.F. Balandrin. American Chemical Society, Washington, DC, p. 96–111.

Spieler, J.M., 1981. World Health Organization, the Special Programme of Research, Development and Research Training in Human Reproduction Task Force on Indigenous Plants for Fertility Regulation. *Korean J. Pharmacogn.,* 12: 94–97.

Suffness, M. and Douros, J., 1982. Current status of the NCI plant and animal product program. *J. Nat. Prod.,* 45: 1–14.

Verpoorte, R., 2000. Pharmacognosy in the new millennium: Leadfinding and biotechnology. *J. Pharm. Pharmacol.,* 52: 253–262.

Vlietinck, A.J. and Vanden Berghe, D.A., 1991. Can ethnopharmacology contribute to the development of antiviral drugs? *J. Ethnopharmacology,* 32: 141–153.

Chapter 6

Anticancer and Cytotoxic Effect of *Mylabris Cichorii* Extract Against Dalton's Lymphoma Ascites-Induced Tumor Model

Akalesh Kumar Verma, Surya Bali Prasad
and Manobjyoti Bordoloi*

ABSTRACT

The present ethnozoological study describes the traditional knowledge related to the use of Mylabris cichorii (Blister beetle) extract in cancer suspected cases by the inhabitants of Karbi Anglong district of Assam, India. The field survey was conducted from March to September 2008 by performing interviews through structured questionnaires with 30 informants, who provided information regarding therapeutic uses of this beetle. The recent study is carried out to evaluate its anticancer potential both in vivo and in vitro. For the recent anticancer studies, the ethanol extracts of *M. cichorii* was prepared and further purified by column

* Corresponding author: E-mail: akhileshverma07@gmail.com

chromatography until the single spot observed on the TLC plate for different fraction. The different single purified fraction was tested for its anticancer activity. The preliminary result showed that out of different fraction one fraction (Fr 03) has good antitumor activity in Dalton' lymphoma bearing mice with 65±1.2 per cent ILS (increase in life span) at 20mg/kg body weight/day (i.p). *In vitro* cytotoxicity result showed 50±4 per cent cell death on the cancer cells at the dose of 20μg/ml. This work will be helpful in biodiversity conservation in India and also give a clue to investigate bio-active compound responsible for its anticancer activity.

Keywords: *Mylabris cichorii, Anticancer, Dalton's lymphoma, Cytotoxicity.*

Introduction

North-Eastern region of India is comprised of rich floral and faunal diversity, and many of the animals and plants of this region are traditionally being used by the people of this region for treating different ailments and diseases (Prasad and Rosangkima, 2004; Kakati *et al.*, 2006). Our preliminary survey also revealed that the people of this region frequently use many animals and their products for the treatment of different ailments including cancer suspected cases. *Mylabris cichorii*, commonly known as blister beetle and locally as "Helicaptor pok" is one of them whose water extract is used by the indigenous people of Karbi Anglong and North-Cachar Hills district of Assam, India, against cancer suspected cases. The size of *M. cichorii* ranges from 0.5 to 2 inches, with black, brown, or gray body colors. Adults have spots or stripes of alternate yellow and black color. It belongs to the class Insecta, order Coleoptera, family Meloidae. *M. cichorii* have been reported to be one of the oldest treatments for cancer in recorded history. For more than 2000 years, blister beetles in powdered or tincture form have been used medicinally in Europe, China and elsewhere (Nikbakhtzadeh and Ebrahimi, 2007). It is still employed in traditional Chinese medicine, and interest in the clinical use of cantharidin as an antitumor agent has re-emerged following the reports that cantharidin produce cytotoxic effects in a number of human tumor cell lines and primary tumor cells (McCluskey and Sakoff, 2001; Wang *et al.*, 2000).

Ingredients sourced from wild plants and animals are not only used in traditional medicines, but are also increasingly valued as raw materials in the preparation of modern medicines and herbal preparations. Animals and products derived from different organs of their bodies have constituted part of the inventory of medicinal substances used in various human cultures since ancient times (Adeola, 1992; Lev, 2003), and such uses still exist in traditional medicine. The healing of human ailments by using therapeutics based on medicines obtained from animals or ultimately derived from them is known as zootherapy. The phenomenon of zootherapy is marked both by a broad geographical distribution and very deep historical origins. The animal-derived remedies constitute an integral part of folk medicine in many parts of the world, particularly for people with limited or no access to mainstream medical services (Lev, 2006; Jamir and Lal, 2005; Alves and Rosa, 2007). Zootherapy constitutes an important alternative among many other known therapies practiced worldwide. Wild and domestic animals and their by-products *e.g.*, hooves, skins, bones, feathers, tusks etc. form important ingredients in the preparation of curative, protective and preventive medicine (Anageletti, 1992). For example, in India, nearly 15- 20 percent of the Ayurvedic medicine is based on animal-derived substances (Unnikrishnan, 1998).

Based on the literature survey and primary information received from the elders and local herbal practitioners through interview, the present study was undertaken to evaluate the possible anticancer activity of *M. cichorii* fractions against murine tumor model.

Materials and Methods

Animals and Tumor Model

Inbred Swiss albino mice (27-30 g) were maintained in under conventional laboratory conditions at room temperature ($20 \pm 2\,°C$) with free access to food pellets (Amrut Laboratory, New Delhi) and water *ad libitum*. Ascites Dalton's lymphoma tumor is being maintained *in vivo* in 10- 12 weeks old mice by serial intraperitoneal (i.p.) transplantations of 1×10^7 viable tumor cells per animal (0.25 ml phosphate buffered saline, PBS, 0.15 M NaCl, 0.01 M sodium phosphate buffer, pH 7.4). Tumor-transplanted hosts usually survived for 19- 21 days. The use of animals in the present study is as per the ethical norms and has been cleared by institutional ethical committee of North Eastern Hill University, Shillong.

Preparation of *Mylabris cichorii* Fractions

M. cichorii (beetles) were collected from different locations of Karbi Anglong district of Assam (India). The whole body of beetles was washed with 90 per cent alcohol and dried in an oven under 40- 45 °C. The dried material was grounded into a powder form using sterilized mortar and pestle. The beetles extract was prepared following the method described by Saxena *et al.* (2007) with slight modifications. The powdered material was hydrolysed using 6 N HCl at 125 °C for 4 h for resolving biomatrix and to set the bound compound free, if any. The powdered material was then extracted with distilled ethanol at room temperature for 3 days. The extract-solvent mixture was filtered using Whatman's No 1 filter paper and dried in a rotary evaporator at 38-40 °C. Ethanol extract was then further purified by column chromatography. Finally the active fractions were isolated by Preparatory thin layer chromatography (PTLC) and tested for anticancer activity.

Antitumor Activity Study

For antitumor studies different fractions were dissolved in phosphate buffered saline (PBS, pH 7.4) and their antitumor activity was determined following the method described by Ahluwalia *et al.* (1984). Tumor cells were transplanted intraperitoneally in 10- 12 weeks old male mice and the day of tumor transplantation was taken as day zero. The tumor-transplanted animals were divided into seven groups with 10 mice in each group. Treatment with different doses (10, 20, 30 and 40 mg/kg body weight/day) of *M. cichorii* fractions and cisplatin (4 mg/kg body weight/day) were given for five consecutive days starting from day 6 of tumor transplantation, and the host survival patterns were recorded. The group of tumor-bearing control mice received same volume of extract vehicle (PBS) alone. The cisplatin treated group served as positive control. The deaths of animals, if any and tumor volume of animals in different treatment groups were recorded daily. The antitumor efficacy was reported in percentage of average increase in life span (per cent ILS) calculated using the formula:

$$\left(\frac{T}{C} \times 100 \right) - 100$$

where,

T and C are the mean survival days of treated and control groups of mice respectively. The most potent fraction, showing highest antitumor activity (65.3 per cent ILS) was selected for further viability studies.

Short-Term Cell Culture and Viability Test

Thymocytes and DL cells (1×10^6cells/ml) were cultured in RPMI-1640 and DMEM supplemented with 10 per cent sheep serum, penicillin (100 IU/ml), and streptomycin (100 µg/ml) in a humidified atmosphere of 5 per cent CO_2 at 37°C until confluent. The cells were dissociated with 0.2 per cent trypsin, 0.02 per cent EDTA in PBS. To analyse the comparative cytotoxicity of the extract on cancer and normal cells, cells were incubated with different doses (10, 20, 30 and 40 µg/ml) of active fraction (Fr 03) of *M. cichorii* and cisplatin. After 3 hours of incubation, cell viability was checked by trypan blue exclusion test (Talwar, 1974a). An aliquot of the cell suspension was mixed with an equal volume of trypan blue (0.4 per cent in PBS) and incubated for 10 min. Number of viable and dead cells were determined with a Neubauer haemocytometer under light microscope. The percentage of viability was calculated using the formula:

$$\% \text{ Viability} = \frac{\text{Total viable cells of treated}}{\text{Total viable cells of control}} \times 100\%$$

Statistical Analysis

Results were Mean ± SE. Significant differences between control and different treatments were calculated using Student's *t*-test. Number of replicates (N) = 5, *p*-

Values of ≤ 0.05 were considered significant.

Results

Percentage Yield of the Extracts

The percentage yield of different fractions of Fr 01, Fr 02 and Fr 03 were 3, 6 and 7 per cent, respectively.

Antitumor Activity of the Extracts

The various fractions and survivability of the hosts in different experimental groups have been shown in Figure 6.1. In the present study, among three different fractions of *M. cichorii* Fr 03 showed

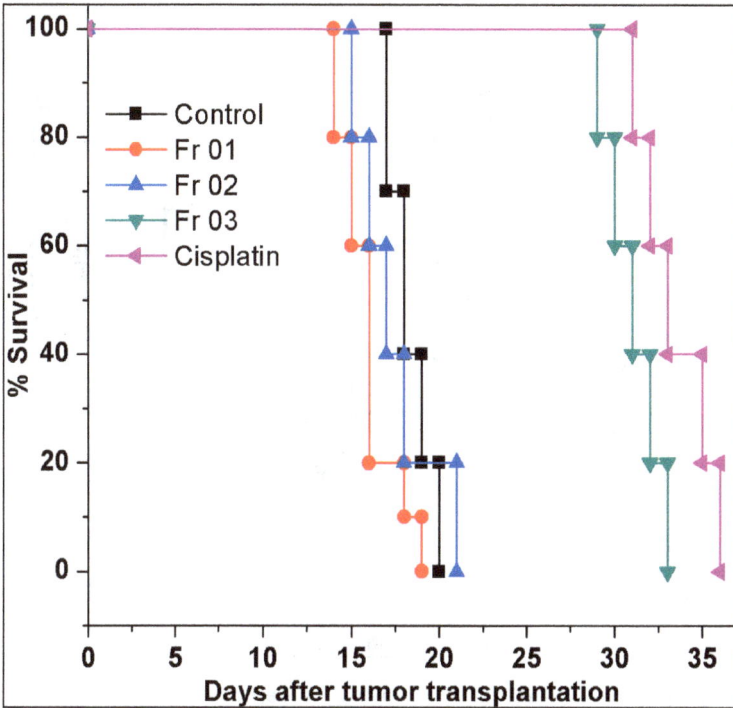

Figure 6.1: Graph Showing the Survival Pattern of the Tumor Bearing Mice After Treatments with Different Fractions.

Control (tumor-bearing mice without treatment). Cisplatin is served as reference drug. Results are expressed as a mean of 3 independent experimental sets.

much better antitumor activity against Dalton's ascites lymphoma with ILS (Figure 6.2) of 65 per cent at 20 mg/kg/day. Cisplatin, a positive control showed 80 per cent ILS at a dose of 4 mg/kg/day. As compared to tumor-bearing control, significant decrease in the tumor volume (Figure 6.3) was observed after treatment with Fr 03 fractions of *M. cichorii,* while other fractions did not show significant changes.

Cell Viability

In the dose-response curve obtained from the trypan blue dye exclusion test, a significant decrease ($p \leq 0.05$) in the percentage of cell viability was observed after the cells were treated with Fr 03

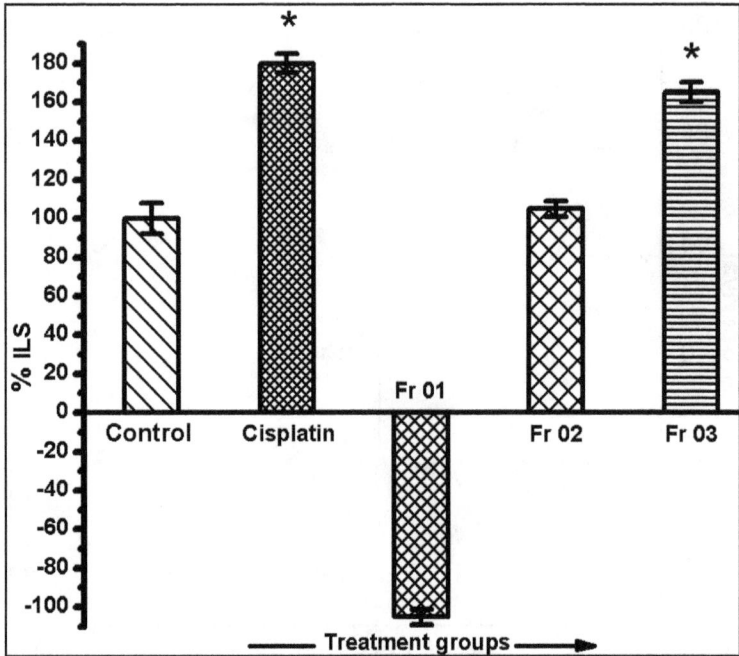

Figure 6.2: Histogram Showing Percentage Increase in Life Span (per cent ILS) of Tumor-Bearing Mice Treated with Different Isolated Fractions

Cisplatin is used as reference drug. Values are mean ± S.E. Student's t-test; as compared to the respective control values, n = 3, *P ≤ 0.05

fraction and cisplatin at 10, 20, 30 and 40 µg/ml for 3 h. Fr 03 fraction showed more cytotoxic effect towards DL cells as compared to thymocytes of normal animals. The IC_{50} values of Fr 03 fraction determined by the trypan blue exclusion test were observed to be 20 and 67 µg/ml in DL cells and thymocytes respectively.

Discussion

In the studies related with the assessment of antitumor activity, ascites Dalton's lymphoma has been commonly used as an important murine experimental tumor model (Prasad and Giri, 1994; Rosangkima *et al.*, 2008). The result of present studies indicate that out of three different pure isolated fractions of *M. cichorii*, Fr 03 fraction showed comparatively better antitumor activity against

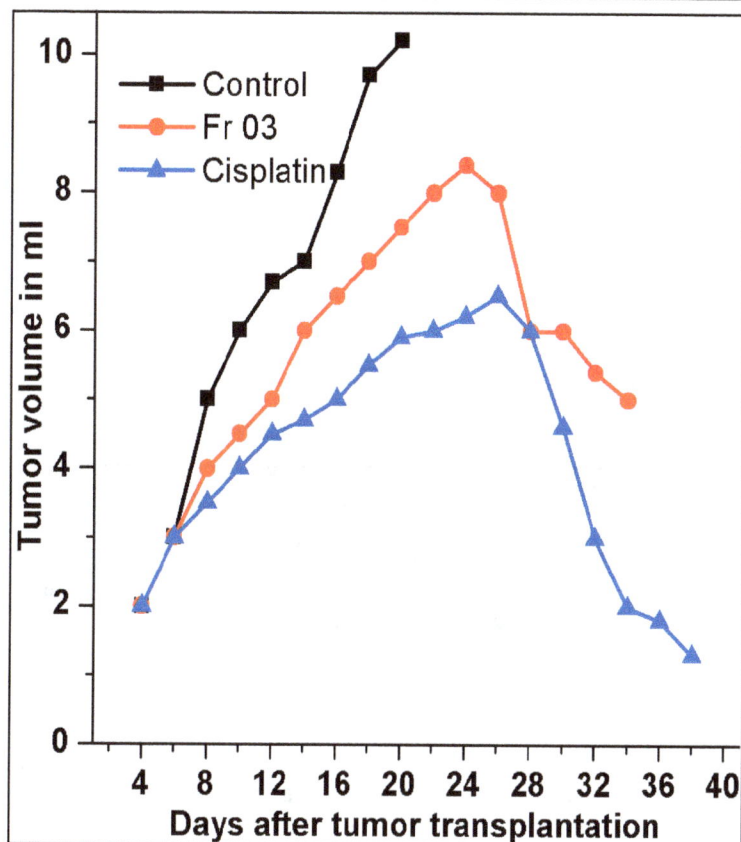

Figure 6.3: Graphs Showing Changes in the Tumor Volume (ml) in Control, Cisplatin (2 mg/kg/day) and Fr 03 Fraction (20 mg/kg/day) Treated Group of Mice. The results are expressed as mean ± S.E, n = 3.

ascites Dalton's lymphoma in mice. Changes in tumor volume in treated group showed significant decrease with longer survivability of the hosts as compared to control group (Figure 6.1 and 6.3). From the data, it may be suggested that 20 mg/kg/day may caused to induce cancer cell death, while higher doses did not show significant increase in survivability of the hosts which could be due to some toxic effect of the extract (Bodenner *et al.,* 1986 and Huan *et al.,* 2006). The result of present *in vitro* cell viability studies also revealed that

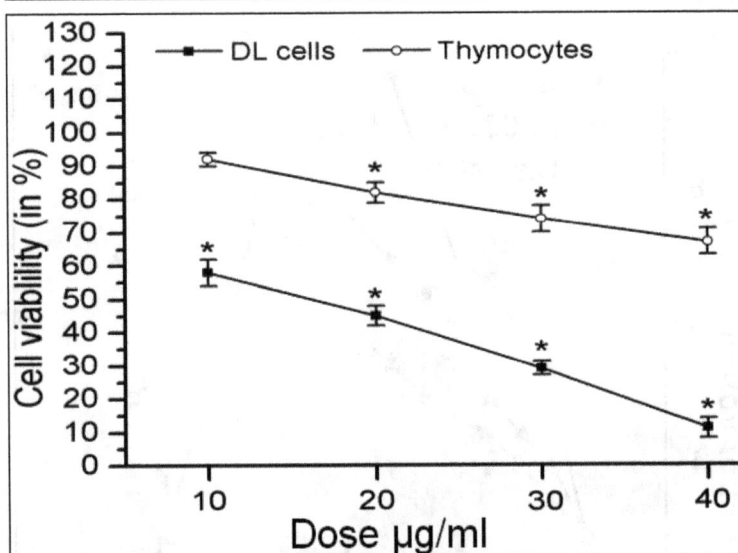

Figure 6.4: Graph Showing the Cytotoxic Effect of Fr 03 Fraction in DL Cells and Thymocytes *in vitro*. Values are mean ± S.E. Student's t-test; as compared to the respective control values, n = 5, *P ≤ 0.05.

Fr 03 fraction at a concentration lower than 20μg/ml showed cytotoxic effect on tumor cells having no significant damaging effect on normal cells, while concentrations higher than 20 μg/ml may have a damaging effect on normal cells also such as thymocytes.

In conclusion, out of different fractions of *Mylabris cichorii* studied Fr 03 fraction showed better antitumor activity against murine ascites Dalton's lymphoma. It is also showed comparatively higher cytotoxic effect towards DL cells as compared to thymocytes of normal animals. The exact mechanism(s) involved in the antitumor activity of *Mylabris cichorii* may not be clear at present. Therefore, further investigation is necessary to determine a detailed mechanism(s) involved.

Acknowledgement

We acknowledge the University Grant Commission, New Delhi (India) for providing Research fellowship in science for meritorious students (F-4-1/2006(BSR)/5-120/2007(BSR), Dated 20th January'09. The author is also thankful to the Director of NEIST, Jorhat for providing the facility for isolation of active fractions in Natural Product Chemistry Division.

References

Adeola, M.O., 1992. Importance of wild Animals and their parts in the culture, religious festivals, and traditional medicine of Nigeria. *Environmental Conservation*, 19: 125–134.

Ahluwalia, G.S., Jayaram, H.N., Plowhan, J.P., Cooney, D.A. and Johns, D.G., 1984. Studies on the mechanism of activity of 2-b-Dribofuranosyl thiazol-4-carboxamide. *Biochemistry and Pharmacology*, 33: 1195–1203.

Alves, R.R., Rosa, I.L. and Santana, G.G., 2007. The role of animal-derived remedies as complementary medicine in Brazil. *Biosciences*, 57: 949–955.

Anageletti, L.R., Agrimi, U., Curia, C., French, D. and Mariani-Costantini, R., 1992. Healing rituals and sacred serpents. *Lancet.*, 340: 223–225.

Bodenner, D.L., Dedon, P.C., Keng, P.C., Katz, J.C. and Borch, R.F., 1986. Selective protection against cisplatin induced toxicity in kidney, gut and bone marrow by DDTC. *Cancer Research*, 46: 2751–2755.

Huan, S.K., Lee, H.H., Liu, D.Z., Wu, C.C. and Wang, C.C., 2006. Cantharidin-induced cytotoxicity and cyclooxygenase 2 expression in human bladder carcinoma cell line. *Toxicology*, 223: 136–143.

Huh, J.E., Kang, K.S., Chae, C., Kim, H.M., Ahn, K.S. and Kim, S.H., 2004. Roles of p38 and JNK mitogen-activated protein kinase pathways during cantharidin-induced apoptosis in U937 cells. *Biochem. Pharmacol.*, 67: 1811–1818.

Jamir, N.S. and Lal, P., 2005. Ethnozoological practices among Naga tribes. *Indian Journal of Traditional Knowledge*, 4: 1100–1104.

Kakati, L.N., Bendang, Ao. and Doulo, V., 2006. Indigenous knowledge of zootherapeutic use of vertebrate origin by the Ao Tribe of Nagaland. *Journal of Human Ecology*, 19: 3163–3167.

Lev, E., 2003. Traditional healing with animals (zootherapy): Medieval to present-day Levantine practice. *Journal of Ethno-pharmacology*, 86: 107–118.

Lev, E., 2006. Healing with animals in the Levant from the 10th to the 18th century. *Journal of Ethnobiology and Ethno-medicine,* 2: 11–26.

McCluskey, A. and Sakoff, J.A., 2001. Small molecule inhibitors of serine/threonine protein phosphatases. *Mini Review. Med. Chem.,* 1: 43–55.

Nikbakhtzadeh, M.R. and Ebrahimi, B., 2007. Detection of cantharidin-related compounds in *Mylabris impressa* (Coleoptera: Meloidae). *J. Venom. Anim. Toxins incl. Trop. Dis.,* 13(3): 686–693.

Prasad, S.B. and Giri, A., 1994. Antitumour effect of cisplatin against murine ascites Dalton's lymphoma. *Indian Journal of Experimental Biology,* 32: 155–162.

Rosangkima, G. and Prasad, S.P., 2004. Antitumor activity of some plants from Meghalaya and Mizoram against murine ascites Dalton's lymphoma. *Indian Journal of Experimental Biology,* 42: 981–988.

Rosangkima, G., Rongpi, T. and Prasad, S.B., 2008. Role of glutathione and glutathione-related enzymes in the antitumor activity of *Dillenia pentagyna* in Dalton's lymphoma-bearing mice. *International Journal of Cancer Research,* 4(3): 92–102.

Saxena, M., Srivastava, S.K., Faridi, U., Mishra, R., Gupta, M.M., Darokar, M.P., Singh, Digvijai., luqman, S. and Khanuja, S.P.S., 2007. Cytotoxic Agents from *Terminalia arjuna. Planta Medica,* 73: 1486–1490.

Talwar, G.P., 1974. *Hand Book of Practical Immunology.* National Book Trust, New Delhi, pp. 329.

Unnikrishnan, P.M., 1998. Animals in Ayurveda. *Amruth,* pp. 1–15.

Wang, C.C., Wu, C.H., Hsieh, K.J., Yen, K.Y., Yang, L.L., 2000. Cytotoxic effects of cantharidin on the growth of normal and carcinoma cells. *Toxicology,* 147: 77–87.

Zhang, W.D., Zhao, H.R., Yan, Y., Wang, X.H., Zong, Z.H., Liu, Y., 2005. Apoptosis induced by cantharidin in human pulmonary carcinoma cells A549 and its molecular mechanisms. *Zhonghua Zhong Liu Za Zhi,* 27: 330–334.

Chapter 7

Prevention of Biopiracy of Traditional Knowledge in North East India: Some Loopholes and Remedies

Rajib Lochan Borah *

ABSTRACT

Biopiracy of traditional knowledge involves gaining of illegal benefits by using traditional knowledge of any indigenous community or tribe without their consent. Such activities may even result into copyrights and patents. India has fought many historical battles with powerful nations like the U.S. for patents, for protection of its resources and indigenous knowledge. North East India has vast tribal diversity with a vast traditional knowledge yet to be protected by patent or by copyright. But, the knowledge holding traditional medicine men and women are mostly illiterate. Though the government formed many rules and acts, yet in practice, results are not satisfactory. There are loopholes by which bio-pirates can easily escape the hand of law. So, the paper wants to highlight some possible efforts

* Corresponding author: E-mail: rajibl_borah@yahoo.co.in

to be made, to bring the indigenous medicine, men and women to the forefront by giving them the opportunity of benefit sharing, to bridge the gap between knowledge holders and laboratories, to involve the traditional knowledge holders in. publishing of scientific literatures. Only by linking the medicine men and women to the scientific community, we can prevent biopiracy.

Keywords: Biopiracy, North-East India.

Introduction

Bio-piracy refers to the use of intellectual property systems to legitimize the exclusive ownership and control over biological resources and biological products and processes that have been used over centuries in non-industrialized culture. Patenting, something which is already known to people, but not published or proven scientifically. Biopiracy denotes a practice where commercial firms (of developed nations) patent and claim ownership of TK (of developing nations).Publication of literatures (research papers, books) of TK, without the consent of the knowledge providing community (the main concern of Biopiracy in NORTH EAST INDIA). Taking copyright of something that is orally run in the community but not published in written form. MNCs of the developed world who are instrumental in appropriation of the precious bio-resources and the allied TK of the South, without the prior informed consent (PIC) of the legitimate holders of such resources, thereby bypassing any benefit-sharing arrangement with them. The mounting profits emanating from the commercial exploitation of these prized resources of the South accrue solely to the corporate giants of the developed world, courtesy the exclusive patent protection obtained by suppressing the true source of the relevant bio-resource and/or TK. The country of origin of the concerned bio-resources and/or TK however remains deprived of its legitimate share of the profits generated out of them. Such unscrupulous business practices of the MNCs have come to be known as biopiracy.

Loopholes of Biopiracy of TK

Patent Itself

Patent offices in some countries require only that the patented bits be novel in their own country, and completely ignored the knowledge of other nations. Countries like India that are rich in biodiversity and traditional knowledge are seeking to end this biopiracy. Several cases of biopiracy have come to the fore in the past as well. The patents granted for instance, on wound-healing properties of 'haldi', or on anti-diabetic properties of `karela', `jamun', and brinjal, are some such examples of blatant biopiracy of Indian TK that have generated considerable hue and cry in not-too-distant past.

Illiteracy of Community

Tribal people generally are unaware, how they are exploited, as they are illiterate.

Oral Knowledge

As tribal people are illiterate, they pass the knowledge to their new generations verbally. Thus no one ever finds the knowledge in written form. This is a serious problem, as the scientific community always recognizes scientific valid publications.

Absence of Scientific Proof

Though traditional medicines are administered for years in tribal societies, scientific community cannot accept the knowledge directly without chemical tests. At this point the western developed countries can take the advantage over developing countries.

Lack of Awareness

General public is not aware that something has been taken away from them, or it becomes too late when they realize it. Example can be given of the patent of Basmati, turmeric and Neem etc. When these are patented, then only we had to challenge and struggle.

No Knowledge of IPR

Even the educated people are unaware of the rules, regulations and laws attached with IPR, copyright and patents. Thus, only an exclusive group of people (lawyers, scientists) have to fight for the cause.

Unethical Bioprospecting Contracts

The ethical debate has sparked a new branch of international patent and trade law. Bioprospecting contracts lay down the rules, between researchers and countries, of benefit sharing and can bring royalties to lesser-developed countries. However, the fairness of these contracts has been a subject of debate. Unethical bioprospecting contracts (as distinct from ethical ones) can be viewed as a new form of biopiracy.

The Protection of Plant Varieties and Farmers' Rights Act of 2001 (PPVFR Act)

It complies with the World Trade Organization's agreement on trade-related aspects of intellectual property rights. The act protects the rights of plant breeders, farmers and researchers over plant varieties.PPVFR Act, which allows anyone conducting research free access without prior informed consent to any genetic resource, including varieties protected by plant breeders' rights. The PPVFR Act does not differentiate the nationalities of people or organisations accessing Indian genetic resources, including varieties protected by plant breeders' rights, for breeding new varieties. The only exception is the need for prior informed consent for repeated use of such a protected variety as a parental line for the commercial production of a new variety.These mean that non-Indian entities can freely access plant genetic resources and associated knowledge for use in breeding or for bio-surveys within India.Secondly, having freely accessed the genetic resources of choice to develop breeding lines or new varieties or nothing, seeds of this material can be taken out in different pretexts as 'exports'.

Government and Official Efforts so far

In response to biopiracy threats faced in cases of turmeric, neem and basmati rice, Government of India has been translating and publishing ancient manuscripts containing old remedies in electronic form, and in 2001 the TKDL was set up as repository formulations of various systems Indian medicines as *Ayurveda*, *Unani* and *Siddha*. The texts are being recorded from *Sanskrit*, *Urdu*, *Persian* and *Arabic*; made available to patent offices in English, German, French, Japanese and Spanish. The aim is to protect India's heritage from being exploited by foreign companies. Hundreds of *Yoga* poses are also kept in the collection. The project has been criticized by a spokesman for the pharmaceutical industry as "a solution in search of a problem". The library has also signed agreements with leading international *patent offices* such as *European Patent Office* (EPO), *United Kingdom Trademark and Patent Office* (UKPTO) and the *United States*

Patent and Trademark Office to protect TK from biopiracy as it allows *patent examiners* at International Patent Offices to access TKDL databases for patent search and examinations purposes.

At the Earth Summit held in 1992, the Convention on Biological Diversity (CBD) was concluded, to which India is a party. The basic objectives of the CBD are: conservation, sustainable use of biological diversity and equitable sharing of benefits arising from the use of biodiversity. It further mandates the signatories to it to respect, preserve and maintain knowledge, innovations and practices of indigenous and local communities and encourage the equitable sharing of benefits arising from the utilization of such knowledge, innovations and practices. As a legally binding treaty, the CBD can be expected to have some influence on these issues.

To ensure that there are legal mechanisms in place to ensure that this knowledge is not freely appropriated, the Indian government is in the process of finalizing a law titled the Biological Diversity Bill. The bill contains various provisions for regulating access to biological resources, patent claims, and indigenous knowledge. This bill is a beginning, though inadequate.

Challenge Patent

Applications for patents should mandatorily disclose the source of origin of the biological resource knowledge pertaining to it, so as to facilitate benefit sharing with the originators of the knowledge and resource. While the neem victory should certainly be regarded as a breakthrough for the legitimate rights of the holders of TK - not only in India but throughout the South - the time, money and effort involved in the whole process of revocation (which in this particular case lasted for ten long years) does stand out as a cause of concern in following the same path in case of other similar cases of biopiracy in future. In fact, the way biopiracy. has been assuming endemic proportions, the transaction costs involved in getting biopiracy patents examined and revoked in foreign patent offices on a case-by-case basis could turn out to be prohibitive for a developing country like India. Hence, the necessity for an internationally enforceable legal regime, which can ensure an effective protection for the rights of communities on their TK-based biological resources by prohibiting the unscrupulous biopiracy practices of the western MNCs.

National Knowledge Commission

Action programe initiated by APJ Abdul Kalam and M.S. Swaminathan on october 2, 2005. Report was finalized on October 2, 2008. According to its directives village knowledge centres will be established. These centers will collect and preserve traditional knowledge.

Geographical Indications Bill

India has developed a multi-pronged approach to tackle IPR disputes. Under this, a product will be defined by a geographical area where it is traditionally found. If the Bill becomes an Act, it will also evolve product standards, provide cataloguing and classification and enforce discipline. However, the problem arises is that the WTO does not recongnise geographical indications for products other than wines and spirit. Though India has put in a proposal to include other products like Kanjeevaram silk, Alphonso mango and Darjeeling tea under this, the WTO is yet to respond to it. Once this legislation is passed, we can state that basmati is from a particular geographical area and another country cannot patent it.

Plant Varieties Bill

This will enable farmers or plant breeders to register their own innovations and traditional knowledge, so they cannot be patented elsewhere. This will definitely ease a lot of farmer's problems.

The Checklist of Issues Submitted by India and Allies to the TRIPS Council in March 2004

Disclosure of source and country of origin of the biological resource and of the traditional knowledge used in the invention:

☆ How would an obligation for disclosure of country and source of origin of biological resource and associated traditional knowledge used in an invention help in better examination of patents and in preventing cases of bad patents?

☆ What is the meaning of disclosure of source and country of origin of biological resource and of the traditional knowledge used in the invention?

☆ What would be the legal effect of wrongful disclosure or non-disclosure?

☆ On whom should the burden of proof lie?

☆ In what manner should the proposed obligation of disclosure of source and country of origin and associated traditional knowledge be introduced in the TRIPS Agreement?

Disclosure of evidence of prior informed consent under the relevant national regime:

☆ How would furnishing the above evidence facilitate achieving the objectives of the CBD of ensuring prior informed consent and harmonious relationship between the CBD and the TRIPS Agreement? Could contractual arrangements for ensuring prior informed consent and benefit-sharing suffice to achieve the objectives of the CBD in this regard?

☆ How should the evidence of prior informed consent through approval of authorities under the relevant national regime be provided for?

☆ What should be the nature of obligation on the patent applicant that should satisfy the requirement of prior informed consent?

☆ What should be the obligation if there is no national regime in the country of origin?

☆ What should be the legal effect of not providing evidence of prior informed consent through approval of authorities under the relevant national regime?

Disclosure of evidence of benefit sharing under the relevant national regime:

☆ What should be the meaning of evidence of benefit sharing under the relevant national regime?

☆ When is this evidence to be introduced by the patent applicant?

☆ What should be the obligation if there is no relevant national regime in the country of origin?

☆ What should be the legal effect of not providing evidence of fair and equitable benefit sharing under the relevant national regime?

Blacklisting

Blacklisting a Biopirate

Someone who is branded as a "biopirate" will suffer from a bad reputation. Someone who gains a reputation for evading access restrictions, or for being hard to deal with, may find it increasingly difficult to find doors open for further research. A company that is associated with biopiracy may end up with weak patents, be exposed to equitable claims for profit-sharing, lose sources of supply, face the prospect of consumer and government boycotts, barriers to importation of biotechnology products, and other loss of market share, and may face financial penalties.

Permission Procedure

Permission to enter an area with tribal community should be made transparent to the community itself.

License Cancellation

If an industry is found to be indulged in biopiracy, it should be punished by procedures such as license cancellation.

Fine as Punishment

A sum of money may be fined and thes money may be utilized for welfare of the knowledge holfing community.

Other Efforts

Peoples Biodiversity Register

NGOs and institutions in India are attempting to document the, knowledge skills and techniques of local communities related to biological resources through the Community (or People's) Biodiversity Register, in the belief that such documentation would be a deterrent to biopiracy; as well as for instilling a greater sense of pride among local communities over the knowledge they possess. The Register processes documents of community and individual knowledge of occurrence, practices of propagation, sustainable harvests and conservation, as well as economic uses of biodiversity resources. All information accumulated in the Register can be used or distributed only with the knowledge and consent of the local community, so that it is in a position to refuse access to the register and to set conditions under which access would be allowed. The community, while consenting, can charge fees for access to the Register and collection of biological resources. Decisions on how to disburse the funds are to be made through village community meetings. These registers are expected to function as tools to establish claims of individuals and communities over knowledge and uses of biodiversity resources, and to bring to them an equitable share of benefits flowing from the use of such knowledge and resources. This, however, can be achieved only when legal mechanisms of control over the register are put in place, which is not yet the case.

Benefit Sharing

Examples can be taken from the kani tribes benefit sharing with TBGRI: The Kani tribals in Thiruvannathapuram district, Kerala, claim that one can live for days together without food, and still be able to perform rigorous physical work, by eating a few fruits of a plant called Aarogyapaccha everyday. The term means the greener of health, the one that gives very good health and vitality. Scientists from the Tropical Botanic Garden Research Institute (TBGRI), learnt about the use of the plant from the Kanis and conducted detailed investigations on the same. Study of the leaves of the plant revealed it had anti-stress, anti-hepatotoxic and immunodulatory/immunorestorative properties. Eventually, the drug Jeevani was formulated by TBGRI with Aarogyapaccha and three other medicinal plants as ingredients. Thereafter, a license to manufacture Jeevani was given to Arya Vaidya Pharmacy, Coimbatore (AVP) in 1995, for a period of seven years, for a fee of Rs. 10 lakhs. TBGRI decided that the Kani tribals would receive fifty per cent of the licence fee, as well as 50 per cent of the royalty obtained by TBGRI on sale of the drug.

Sacred Grooves

Sacred grooves are traditionally protected forest patches managed by village community on socioreligious grounds. The knowledge of the biodiversity of the sacred groove is kept by the community. This knowledge can be tapped by paying a benefit sharing money or taking permission from the community.

Best Ways Possible

1. Bridge gap between tribal community and learned community.
2. Knowledge of scientific literature to the community.

3. Encourage to write down the TK for future.

4. Formation of peoples biodiversity register.

5. Invitation of medicine men and women to seminars and conferences.

6. NGO involvement.

7. Consent letter should be made compulsory, if someone writes any literature on TK.

8. Educate the people.

9. Scientific evaluation of the medicines prepared traditionally.

10. Cancelled patents on natural product inventions: Patents on natural product inventions are subject to attack unless all public knowledge about the species in question and its use are fully disclosed. Organizations in the bioresource-rich but economically poor countries of the developing world have demonstrated a willingness to attack natural product patents on the basis of TK turmeric, *basmati, neem* etc.

11. Corporations and research facilities should not accept illegally-obtained biological materials from collectors.

12. *Denial of access to samples:* As a practical matter, if a collector does not agree to provide an equitable share of benefits, in advance, to the source of biological samples, the collector may well be denied access to the samples. Simply put, the possibilities for fieldwork will dry up.

13. *Other Legal penalties:* Finally, the ultimate legal sanction — criminal penalties, including jail — may apply. It is not uncommon for hunters to be jailed for poaching or trespassing. In the biodiversity prospecting context, there is at least one instance of a researcher who was temporarily detained in Australia for unauthorized collecting of plant materials. Collection of biological materials without a benefit-sharing agreement is likely to find its way into the list of criminal violations in some countries, so that biopiracy could result in a jail sentence.

Conclusion

Policies are easier to develop than monitor.

References

Anonymous, 2006. Intellectual property rights. *Yojana,* February.

Bhakat, R.K., 2009. Protecting a Sacred Grove. *Yojana,* August.

Katti, V., 2005. The new patent regime, *Kurukshetra,* May.

Parnava, H.K., 2009. Role of NGO s in Environment Protection. *Science Reporter,* October.

Rao, R., 2004. Traditional Knowledge Digital Library. *Kurukshetra,* September.

Sanskaran, P.N., 2005. National Knowledge Commission. *Kurukshetra,* August.

Udgaonkar, S., 2004. The recording of traditional knowledge, 413–419, *Current Science,* Vol–82, No–4, 25 February.

Chapter 8

A Brief Review on Some of the Medicinal Properties of Tea (*Camellia sinensis* L.)

Nizara Bhattacharyya

ABSTRACT

Tea (*Camellia sinensis* L.) a non-alcoholic beverage is also recognized as an herbal medicine from ancient time. In modern science, the pharmacological/biological effects of tea drinking may include antimicrobial, diuretic, antipyretic, immune stimulating, as well as cardiovascular diseases and cancer prevention. Majority of the biological studies with tea have been carried out with green tea and reports on biological and pharmacological properties of black tea are scanty. It is also documented that some of the tea constituents are also used for beauty and health purposes. Tea- therapy is one of the important components of dietotherapy. Tea has been suggested to remove tiresomeness of the body and fatigue of the mind and the pre-game tension and post-game fatigue in the case of players and sportsman.The composition of tea is very complex and more than 450 components have been identified in tea leaves. Among them, the most important group is tea polyphenols, accounting for about 36 per cent of the dry weight. The major part of green tea polyphenols, commonly known as catechins. In the manufacture of black tea, the flavanols are subjected to oxidation, leading to the formation of theaflavins, thearubigins, etc.in this paper, a brief report is generated on the recent information on the pharmacological and biological properties of green tea and black tea constituents so that tea can be popularized not only as hot and cheapest beverages but also as health beverage.

Keywords: *Chemical composition, Caffeine, Antioxidant, Pharmacology, Anticancer.*

Introduction

Tea is grown in about 30 countries but consumed throughout the world, although at greatly varying levels. It is the most widely used beverage aside from water with a per capita world-wide consumption of approximately 0.12 L per year (Graham, 1991).In contrary to other alcoholic beverages tea drinking cheers, stimulates, and does good to health. Tea, which is manufactured from two leaves and an apical bud of the shrub *Camellia sinensis* var. *assamica,* and other Southern varieties. It is processed differently to give black, green or oolong tea. Green tea constitutes about 20 per cent of, oolong tea 2 per cent of, and the rest is black tea of the total world production. In green tea, which is mostly prepared in Japan and China, the enzymes are inactivated at the outset to prevent oxidation of the leaf polyphenols. In black tea production, oxidation of catechins to yield the right proportion of theaflavins and thearubigins is promoted. Oolong tea is only partly oxidized. In India almost all tea is consumed as black tea. The preparation of the brew in India fluctuates widely from weak infusion to strong extraction, which may be further modified by the addition of milk, sugar, and various spices or spice mixes.

The chemical composition of manufactured tea will vary according to numerous factors such as varietals differences, environmental effect, agricultural management and the methods of processing etc. In unprocessed tea the following chemical compounds are found (Graham, 1983; Stagg and Millin, 1975).

Table 8.1: Chemical Composition of Unprocessed Tea

Compounds	Per cent of Dry Wt.
Carbohydrates	30
Protein	20
Lipid	2
Polyphenols	33-36
Caffeine	4-5
Minerals etc.	7

The carbohydrates comprise the following:

Carbohydrates	Per cent of Dry Wt.
Sugars	4
Starch	2
Pectins	11
Pentosans and Crude fibre (cellulose, lignin)	13

The carbohydrates in tea are of no nutritional or pharmacological significance. The most important chemicals present in tea, which are also of considerable pharmacological significance, are the polyphenols and caffeine. They are present to the extent of 30–35 per cent of the dry weight and determine the quality of the beverage and the amount present depend not only on the genetic make up but also on environmental factors such as climate, light, rainfall, temperature, nutrient availability and leaf age. The varietal difference of total polyphenols and the individual catechins are given in Table 8.2.

Pharmacological Actions

Polyphenols

Tea polyphenols have the properties mopping up oxygen radicals which can be generated in the body by partial reduction of molecular oxygen. The enzymic production of super oxide anion and the presence of superoxide dismutase dedicated to the catalytic removal of this free radical have strengthened the free radical theory of oxygen toxicity. The free radicals can form lipid peroxide which can damage blood vessels and plaque and atherosclerosis of blood vessels (Pant, 1991). Cancer, arthritis, skin wrinkling, and the aging process have been ascribed to superoxide anion causing lipid peroxidation. The polyphenols present in tea can offer a logical chemical explanation for protecting against oxygen toxicity and the hazards of diseases induced by free radicals. The antimutagenic and anticarcinogenic activity is now receiving much scientific attention.

Table 8.2: Variation of Total Polyphenols and Individual Catechins in some TV Clones (per cent dry wt.)

Clone	TP	(-) EGCG	(-) EGC	(-) ECG	(-) EC	(+) C	Total Catechins
TV1	28.09	5.45	3.81	1.85	0.94	0.46	12.51
TV10	26.96	8.56	2.13	1.17	0.64	0.82	13.32
TV17	29.59	5.63	2.08	2.27	1.07	0.92	11.97
TV25	25.99	9.16	2.01	1.30	0.53	0.85	13.85

TP: Total polyphenols; (-)EGCG: Epigallocatechingallate, (-)EGC: Epigallocatechin; (-)ECG: Epicatechingallate; (-)EC: Epicatechin and (+)C: Catechin.

Source: Ann. Sci. Report, Tocklai, 1996-1997.

A list of the physiological actions of tea polyphenols revealed so far is mentioned below (Hara, 1993):

☆ Anti-oxidative Action

    Oxygen radicals

    Plasma lipids

☆ Radio protective Action

☆ Anti- mutagenic Action

    Spontaneous mutations in Bacillus subtilis NIG 1125

    Suppression of mytomycin C-induced micronuclei in mice bone marrow

☆ Anti-tumour Action

    Suppression of the growth of implanted tumour cells

    Suppression of tumorigenesis by inoculated carcinogen

    Suppression of spontaneous breast cancer in mice

☆ Enzyme Inhibitory Action
   Engiotensin I Converting enzyme
   Amylase, Sucrase, maltase
   Glycosyltransferase of Streptococcus mutants
   HIV- reverse Transcriptase
☆ Anti-hypercholesterolemic Action
☆ Anti-hyperglycaemic Action
☆ Fat Reducing Action
☆ Anti-hypertensive Action
☆ Anti-ulcer Action
☆ Anti-bacterial Action
   Food borne pathogenic bacteria
   Phylopathogenic bacteria
   Cariogenic bacteria
   Bacterial exotoxin
☆ Anti-viral Action
   Tobacco mosaic, Influenza
☆ Deodorant Action
☆ Molluskcidal Action

Anti-oxidative Action

It is proved that tea polyphenols (TP) has a strong effect on scavenging super oxygen anion radicals and thus inhibit the formation of other reactive oxygen species. Experiments on rat showed that TP not only reduce the lipoperoxidation (LPO) in serum and cardiac tissue but also promote the activity of SOD in red blood cell. SOD can prevent the damage of super oxygen anion radical from oxygen metabolism. Hence, TP not only can scavenge reactive free radicals and reduce LPO but also improve the immunity by enhancing the antioxidase ability of the body (Wang, 1995),

Anti-tumour Action

The direct studies of evaluating the effectiveness of tea is cancer prevention have been conducted in various experimental model using tumour induced animals by chemicals. The inhibitory effect of tea on chemical induced tumour initiation and promotion has been reported in skin digestive tracts colon lung liver pancreas and mammary gland in rodents. Several epidemiological studies supporting the effects of tea on human cancer, especially the stomach cancer have been reported. Han and Chen (1995) has conducted series of experiments with green tea extract, tea polyphenols and individual catechins and the results indicated that among the catechins, EGCG, EGC and ECG showed stronger effect than other catechins, however the combined effect of tea components are stronger than any individual components for the prevention of carcinogenesis. It was also observed that black tea has the same antimutagenic and anticarcinogenic effects as green tea.

Fat Reducing Action

Akinyanju and Yudkin (1967) administered dried tea at the rate of 1.0 g in 100g diet to rats for 50 days and found a significant decrease in serum cholesrol and triglycerides. They expected a rise in lipid due to caffeine but in fact found that tea produces a fall rather than rise and concluded that tea contains a substance that acts to decrease serum lipids more strongly than caffeine acts to increase them. Black tea extracts are also reported to be effective in reducing blood triglyceride and cholesterol levels (Das *et al.*, 1965).

Anti-bacterial Action

Tea polyphenols also precipitate protein and therefore are astringent. This is the basis of demonstrable beneficial antibacterial activity of tea extracts *in vitro* and their ability to soothe inflamed throat in mild infection. This antibacterial activity has been reported against *Streptococcus mutants* and *Bacillus lactis* the organisms implicated in dental caries. This activity plus the fluoride in tea which can inhibit the glycolytic enzymes responsible for producing organic acids like lactic acid and strengthen tooth enamel through formation of flouroapatite, are considered to bestow on tea an anticariogenic activity. Ishigami and Hara (1993) investigated the minimum inhibitory concentrations (MIC) of the polyphenolic components of green and black tea against 17 strains of food borne pathogenic bacteria and 6 strains of enteric bacteria (Table 8.3). Results showed that tea polyphenols have antibacterial activities against *Bacillus cereus*, 3 *Virio strains*, *Staphylococcus aureus*, *Plesiomonas shigelloides* and *Aeromonas sobria*. Tea polyphenols also have a similar effect against 2 clostridia, *Clostridium perfringens* and *C. botulinum*. The food borne pathogenic bacteria which are vulnerable to TP are largely those responsible for incidences of food poisoning and hence TP may possibly be used to prevent bacterial food borne diseases. Moreover, TP showed virtually no activity against 3 bifidus and 3 lactic acid bacteria which suggested that the habit of tea drinking may improve the intestinal flora

Table 8.3: Percentage of Inhibition of Insoluble Glucan Formation by Tea and TP

Sample	Concentration	Inhibition (per cent)
Instant green tea	1 mg/ml	39.9
	10 mg/ml	79.8
Crude catechins	1 mg/ml	28.0
	10 mg/ml	93.3
(-) Epicatechin (EC)	1 mM	5.5
	10 mM	42.3
(-) Epigallocatechin (EGC)	1 mM	13.1
	10 mM	24.7
(-) Epicatechingallate (ECG)	1 mM	35.5
	10 mM	83.0
(-) Epigallocatechingallate (EGCG)	1 mM	41.6
	10 mM	75.0
Instant black tea	1 mg/ml	30.7
	10 mg/ml	90.5
Crude theaflavins	1 mg/ml	76.8
	10 mg/ml	99.8

Anti-ulcer Action

Marked antibacterial effects were noted and found more in green tea than black tea. Daily administration of tea extract to rats for 7 days significantly reduced the incidence of ulcer as compared to control animals in which ulceration was induced either by administration of aspirin or by subjecting them to cold restraint. Stress reduces the volume of gastric secretion and total acid and increased the peptic activity as compared to control rats. The rats which were administered for 7 days were subjected to cold restraint stress the volume of gastric secretion increased and total acid and peptic activity decreased significantly in comparison to saline treated rats subjected to cold restraint stress. Aspirin significantly increased the total acid and peptic activity as compared to normal rats. When aspirin was administered to rats pretreated with tea extracts, the total acid was decreased significantly as compared to only aspirin group.

Caffeine

Caffeine is a better known and better established pharmacological agent than the tea polyphenols (Gilman *et al.*, 1990). The central nervous system is indiscriminately stimulated by caffeine without subsequent depression. Caffeine increases the motor effects and improves higher function of the brain. It decreases fatigue and increases physical work that can be done (Gaddum, 1959). Caffeine stimulates respiratory, vagal and vasomotor centres in the medulla. It is particularly effective as a respiratory stimulant, without any increase in pulse or blood pressure. It produces peripheral vasodilation, increased circulation in the kidney and brain. It increases the number of active glomeruli in the kidney and has a diuretic action.

Conclusion

Tea has been consumed as a beverage for many centuries mainly it's stimulating and fatigue ameliorating action. The scientists have been unravelling its disease preventing and therapeutic benefits. It does seem that Tea is nature's gift to man, where caffeine and polyphenols supplement and compliment each other beautifully. In the tea cream, they form a complex which reduces the astringency of tea and bitterness of caffeine and release the pharmacologically active principles slowly with beneficial effects on health. Various studies showed that the combined effect of tea polyphenols *i.e.* water extract or crude polyphenols are stronger than any individual component.

Owing to other remarkably inhibitory effects of green tea and black tea extracts or its components on senility, bacteria or toxin infections, hypertension, hyperglycemia, allergic reactions and radiation damage, tea polyphenols was suggested to be used as effective material in the manufacture of health- keeping food products and beverages as well as drugs for prevention and even therapy of diseases.

References

Das, D.N., Ghosh, J.J., Bhattacharya, K.C. and Guha, B.C., 1965. Tea II, Pharmacological aspects. *Indian J. Appl. Chem.*, 28(1): 15–40.

Gaddum, J.H., 1959. *Pharmacology*, 5th edn. Oxford University Press, London, p. 102–104.

Ghoshal, S., Krishna, P.B.N., Raj, K. and Bhaduri, A.P., 1993. Antiamoebic and antibacterial properties of black tea. *Proc. of Int. Sym. on Tea Science and Human Health*, p. 119–124.

Gilman, A.G., Rall, T.W., Nies, A.S. and Taylor, P., 1990. *The Pharmacological Basis of Therapeutics*, 8th edn. Pergamon Press, NY, p. 619–630.

Graham, H.N., 1991. Review of tea consumption, chemical composition, polyphenols chemistry. *Proc. of Int. Sym. on Physiological and Pharmacological Effects of Camellia sinensis* (Tea) N.Y., p. 1–2.

Han, C. and Chen, J., 1995. The screening of active anticarcinogenic ingredients in Tea. *Proc. of International Symposium on Tea Science and Human Health*, p. 39–48.

Hara, Y., 1993. Multifunctional activities of Tea Polyphenols. *Proc. of International Symposium on Tea Science and Human Health*, p. 110–117.

Ishigami, T. and Hara, Y., 1993. Anti-carious and bowel modulating actions of tea. *Proc. of International Symposium on Tea Science and Human Health*, p. 125–132

Maity, S., Vedasiromoni, J.R. and Ganguly, D.K., 1993. Antiulcer effect of hot water extract of black tea (*Camellia sinensis* var *assamica*). *Proc. of International Symposium on Tea Science and Human Health*, p. 133–135.

Mulky, M.J., 1993. Chemistry and Pharmacology of tea. In: *Tea Culture: Processing and Marketing*, (Eds.) Mulky and Sharma, p. 83–96.

Pant, P., 1991. *Beware of Free Radicals*. Financial Express, October 6.

Stagg, G.V. and Millin, D.J., 1975. The nutritional and therapeutic value of tea: A review. *J. Sci. Food Agric.*, 26: 1439–1459.

Wang, H., 1995. Effect of tea polyphenols on antioxidation and reduction of blood lipid. *Proc. of International Symposium on Tea Science and Human Health*, p. 56–58.

Chapter 9

Isolation of Genomic DNA and its Size Determination of *Karphul* (*Etlingera assamica*)

D. Chowdhury and B.K. Konwar

ABSTRACT

A protocol is described to obtain high molecular weight pure genomic DNA from the aromatic and medicinal plant Etlingera assamica, locally known as Karphul under the Zingiberaceae family. It is a modified CTAB (Cetyl-trimethylammonium bromide) based procedure, produced a good amount of 140 µg pure DNA per g of leaf tissue having $\lambda_{260}/\lambda_{280}$ ratio 2.0. An innovative, easy and less expensive method is developed to determine the genome size of the plant and the size of the genome is found to be 1.2 pg or 1173.6 Mb.

Introduction

Karphul (*Etlingera assamica*) is a wild perennial aromatic and medicinal herb under the family Zingiberaceae. Presently, the genus *Etlingera* is known about 70 species distributing from Himalayas and S.W. China through Myanmar, Thailand, Malaysia, Indonesia, New Guinea and North Queensland (Ved Prakash *et al., 1998*). It grows wild infertile sandy loam to clay loam soils of Central and Upper Assam and also in some other parts of the North Eastern region. It possesses long pseudostem and underground creeping rhizomes. Leaves are petiolate, oblong lanceolate, glabrous with prominent mid rib and acuminate apex. In Assam, the herb is mostly used for its aromatic and medicinal properties. Local people use its rhizome against various illnesses like stomach upset (ground rhizome paste is mixed in lime water and then consumed); rheumatism (rhizome paste used locally); respiratory complains, bronchial catarrh; cooked with rice to act as a tonic for faster post partum recovery of mothers and also consumed with betel nut and leaf as masticatory for its aromatic and refreshing action. The rhizome is also used as a flavouring agent in many food items.

In recent years, the natural population of the herb is decreasing alarmingly due to large-scale deforestation, unsystematic collection and lack of awareness among the common people about its potential. Due to these reasons, the herb has become a threatened plant and localized in certain areas only; and if the same trend continues, the herb will disappear from the region forever within a short time. On the other hand, up until now, no serious and systematic attempt has been made to study the herb. In view of potentiality of the plant to use as food flavour and medicine, a systematic approach is necessary to study its molecular aspect to give a new impetus to the plant. By using molecular biological tools, it is now possible to manipulate the level and composition of pharmaceutical and essential oils in medicinal and aromatic plants. The emerging opportunities for the incorporation of genetic engineering into existing breeding programmes for the full exploitation of the biosynthetic potential of aromatic and medicinal plants would be the new trend of research for days to come. Isolation of pure genomic DNA and its size determination is the prerequisite steps in the application modern genetic engineering technology for the directional improvement of the plant.

Keeping this background in mind the present investigation was carried out to standardize the genomic DNA isolation protocol and determine the genome size of *Etlingera assamica*.

Materials and Methods

Reagents and Chemical Required

☆ Tris-Cl pH 8.0 (1.0 M); EDTA pH 8.0 (0.5 M); NaCl (5.0 M); CTAB (20 per cent); Chloroform: Isoamyl alcohol (24: 1 v/ v); Polyvinylpyrrolidone and β-mercaptoethanol.

☆ Extraction buffer: 100 mM Tris-HCl (pH 8.0), 25 mM EDTA, 1.5 M NaCl, 2.5 per cent CTAB, 0.2 per cent β-mercaptoethanol (v/v) (added immediately before use) and 1 per cent PVP (w/v) (added immediately before use).

☆ High salt TE buffer: 1 M NaCl, 10 mM Tris-HCl (pH 8.0), 1 mM EDTA (pH 8.0).

Plant Samples for DNA Isolation

For the isolation of genomic DNA of *E. assamica*, tender leaves were collected from the plant in the morning, washed with distilled water. The excess water adhering to the leaves was soaked off using tissue paper. They were then packed in polythene bags and stored in darkness at room temperature for the subsequent DNA extraction work.

DNA Isolation Protocol

A modified CTAB-DNA isolation protocol was standardized on the basis of the method used by Dellaporta *et al.* (1983), Doyle and Doyle (1987), Porebski *et al.* (1997) and Khanuja *et al.* (1999). The protocol is presented below:

1. Fresh tender leaves were collected (5 g) from the plant, washed thoroughly with sterile distilled water, blotted the excess water with tissue paper and then surface sterilized with 70 per cent alcohol.

2. The sterilized leaves were cut into smaller pieces.

3. The leaf material was ground into a fine powder treating with liquid nitrogen and using a pre-chilled pestle and mortar.

4. The ground powder was transferred directly into a 25-ml polypropylene tube and added 12 ml freshly prepared pre-warmed (56°C) extraction buffer and mixed gently by several inversions.

5. The sample was incubated at 60°C in a shaking waterbath (100 rpm) for 2 h with occasional mixing to avoid reaggregation of the homogenate.

6. To the homogenate, 12-ml of chloroform: isoamyl alcohol (24:1) was added and mixed gently by inversion for 10 minutes.

7. The extract was centrifuged at 10,000 rpm for 10 minutes with SS34 rotor in Sorval RC-5C centrifuge at 25°C. The supernatant was transferred to a clean 15 ml polypropelene tube and the process (adding of chloroform: isoamyl alcohol, mixing and centrifuge) was repeated twice to clear the aqueous phase.

8. A volume of 3 ml of 5 M NaCl was added to the aqueous phase and was mixed properly without vortexing.

9. Isopropanol was added to the mixture at 0.6 volume and mixed by inversion. The mixture was incubated at room temperature overnight to precipitate the nucleic acid.

10. The sample was centrifuged at 10,000 rpm for 10 minutes at 25°C. The supernatant was poured off and washed the pellet with 80 per cent ethanol and then carefully transferred the pellet in to a clean microfuge tube. The pellet was again washed with 80 per cent ethanol.

11. The pellet was dried in a speed vacuum for 15 minutes and was dissolved in 0.5 ml of high salt TE buffer. RNase 5 µl was added to the solution and then incubated at 37°C for 2 hrs.

12. After incubation, the sample was extracted with equal volume of chloroform: isoamyl alcohol (24:1). The aqueous layer was transferred to a fresh 1.5 ml microfuge tube and was added 2 volumes of cold ethanol.

13. The sample was then centrifuged at 10,000 rpm for 10 minutes at 25°C to precipitate the DNA.

14. The pellet was rinsed with 80 per cent ethanol and then dried in a speed vacuum system (). The pellet was dissolved in 200 µl of TE buffer at room temperature and stored at -20°C.

UV-Spectroscopic Determination of DNA Yield

1. From the stored sample, an aliquot of 10 μl of DNA was transferring into a quartz cuvete and made up the volume to 2 ml with distilled water.

2. The cuvete was placed in the UV-spectrophotometer (Beckman DU® 530 Life Science UV/Vis Spectrophotometer) and absorption was measured at 260 nm and 280 nm.

3. The value between the absorption data at 260 nm and 280 nm was calculated to check the purity of the isolated DNA. A good DNA preparation exhibits a value in between 1.8 - 2.0.

4. DNA concentration was calculated by using the relationship of soluble standard DNA being 1 O.D. at 260 nm = 50 μg/ml.

Agarose Gel Electrophoresis

1. 0.8 per cent agarose gel was prepared in 1 X TAE.

2. 2.0 μl ethidium bromide (10 mg/ml) was added to agarose gel.

3. 10 μl DNA added with 2 μl of Bromophenol blue dye was mixed and loaded in the wells in agarose gel along side the standard Hind III digested λ DNA molecular weight marker for visualization.

4. Electrophoresis was carried out at 60 – 70 V for about 1.5 h.

The electrophoresed gel was exposed to UV light using a UV transilluminator for visualizing the DNA. DNA bends seen were documented by taking photographs with a Gel Doc system (BIO RAB Gel Doc 1000).

Genome Size Determination (Estimating DNA C-values)

The genome size of plant or animal cells can be determined by measuring the DNA content of their nuclei. The first C-values estimated for a few plants were made in the 1950s using tedious chemical extraction methods. The subsequent development of other techniques including Feulgen microdensitometry, flow cytometry and more recently DNA image cytometry, have made estimating DNA amount both easier and faster, such C-value data are now available for almost 5,000 plant species. The genome size varies greatly between species. But the existence and extent of intraspecific variation in genome size is a particularly difficult and controversial field and is currently receiving much attention. Nevertheless, remarkable intraspecific variation has been reported many times from the late sixties to present day (Miksche, 1968; Laurie and Bennett, 1985; Rayburn *et al.,* 1985, 1997; Graham, Nickell and Rayburn, 1994; to quote only the earliest and some more recent references). Moreover, the above-mentioned techniques need sophisticated instruments, which is not always possible in all situations. Therefore, we have tried to develop an easy, accurate and rapid method for determining the genome size by measuring the DNA content of a single somatic cell nucleus.

In this method, the average volume of a cell from the leaf tissue was calculated. The number of cells present in a particular leaf tissue with known volume and weight (less 7–70 per cent cells as intercellular space depending upon the species and the habitat of the particular plant, Turrell, 1934) was calculated. If, the amount of DNA per g fresh tissue, number of cells per g tissue and ploidy level of the plant (by Karyotypic study) were known, the amount of DNA in a single cell can be calculated easily by dividing the amount of DNA present in that tissue by the total number of cells in the same tissue. Thus, the genome size of a particular plant can be worked out as the amount of DNA present in the haploid set of chromosomes in a diploid species.

Cell Volume Determination in Leaf Tissue

For measuring the average volume of a single cell of *E. assamica* leaf tissue, fine transverse and longitudinal sections were made with the help of sharp razor blade and mounted on a slide for observation under the compound microscope (Leica ATC 2000 Model). Initially, 10 X 10x magnification followed by 10 X 40x for measuring the length, breadth, depth and diameter (for round shaped cells) of different cells. By studying the *E. assamica* leaf anatomy, it was found that there were two layers of rectangular shaped epidermis cells on both surfaces of the leaf and in between cylindrical and round shaped mesophyll cells. The average volume of 10 randomly selected cells was worked out with the help of micrometer scale.

Within the leaf, there are lots of intercellular spaces in between the cells. The ratio of intercellular space to total cell volume is species-specific and habitat-specific. It was reported that the intercellular space ranges at a ratio of 70-700: 1000 or 7-70 per cent of the total volume (Turrell, 1934). The distribution of intercellular spaces in the leaves of plant grown under shady condition was lesser than those grown under sunny condition. Since, this perennial herb is mostly grown under shady condition, the intercellular space was considered to be *ca.* 30 per cent of the total volume of the tissue.

Therefore, the mathematical deduction can be considered to be:

Average volume of a single cell = $x \mu^3$

Volume of the tissue = $v \mu^3$

So, total number of cells in the tissue would be $y = v/x$

Now, we know the weight of the tissue having y cells = $w g$

Or, $w g$ tissue contains y cells

So, 1 g tissue contains $(y \times 1/w)$ cells

Now, we know 1 g leaf tissue contains $d pg$ DNA

i.e. $(y \times 1/w)$ cells contain $d pg$ DNA

So, one cell contains $d/(y \times 1/w) pg$ DNA

Since, the species is diploid (from karyotypic study)

So, genome size will be $\frac{1}{2}\{d/(y \times 1/w)\} pg$ DNA

Results and Discussion

The young leaves of *E. assamica* were sampled and processed for DNA isolation and purification by a modified CTAB-DNA isolation procedure on the basis of the methods used by Dellaporta *et al.* (1983), Doyle and Doyle (1987), Porebski *et al.* (1997) and Khanuja *et al.* (1999). CTAB was used as a molecular detergent in the extraction buffer to separate polysaccharides from DNA. PVP was added to remove polyphenols and high NaCl was used in the extraction buffer to remove polysaccharides.

The DNA yield from the *Etlingera assamica* fresh leaves by using the above mentioned procedure was found to be 140 µg per g of leaf tissue.

DNA Quantification and Purity Test by UV Spectroscopy

The purity of the isolated DNA was checked by the ratio $\lambda_{260}/\lambda_{280}$. The OD at λ_{260} and at λ_{280} was recorded with a UV spectroscopy.

$\lambda_{260} = 0.042$ and

$\lambda_{280} = 0.021$

Therefore, the ratio $\lambda_{260}/\lambda_{280} = 0.042/0.021 = 2.0$

The good DNA preparation exhibits a value in between 1.8 - 2.0. Since, the calculated value is 2.0, so, we assumed that the isolated DNA is pure one.

The concentration of isolated DNA was calculated by the formula

At 260 nm, 1 O.D. (optical density) = 50 µg/ml

Therefore,   0.042 O.D.   = (50 x 0.042)µg/ml

= 2.1 µg/ml

Isolated genomic DNA for *E. assamica* was electrophoresed through ethidium bromide stained 0.8 per cent agarose gel along side the standard molecular weight marker λDNA digested with *Hind*III. Photograph of the gel was taken with help of Gel Doc System and presented in the Figure 9.1.

Figure 9.1: Isolated Genomic DNA from *E. assamica* Plant Resolved on 0.8 per cent Agarose Gel.
Lane 1: Hind III digested λDNA, ***Lane 2***: Genomic DNA

Genome Size Determination

The genome size of an organism is the amount of nuclear DNA in its unreplicated gametic nucleus, irrespective of the ploidy level of the taxon. For determining the amount of nuclear DNA of *E. assamica*, a novel method was employed, where average volume of single cell of leaf tissue and other related parameters were calculated by the method described in the Materials and Methods.

Therefore, the average volume of a cell in the leaf tissue
= $8.6 \times 10^4 \ \mu M^3$ (Average of 10 randomly selected cells).

Volume of the leaf tissue   = $1.9 \times 10^{11} \ \mu M^3$

Therefore,

Number of cells present in that tissue
= $1.9 \times 10^{11} \ \mu M^3 / 8.6 \times 10^{11} \ \mu M^3$

= 2.2×10^6

As, there would be 30 per cent of the total volume of the leaf tissue intercellular spaces,

So, the actual number of cells in that tissue
= $2.2 \times 10^6 - 30$ per cent of (2.2×10^6)

= $2.2 \times 10^6 - 6.6 \times 10^5$

= 1.5×10^6

Weight of leaf tissue = 0.025 g

i.e., 0.025 g leaf tissue contains = 1.5×10^6 cells

Therefore, 1.0 g leaf tissue = $(1.5 \times 10^6 \times \dfrac{1.0}{0.025})$ cells

= $(1.5 \times 10^6 \times 40.0)$ cells

= 6.0×10^7 cells

Now, we found that 1.0 g leaf tissue contains
= 140 µg genomic DNA

= $140 \times 10^6 \ pg$ DNA

= $1.4 \times 10^8 \ pg$ DNA

Therefore, 6.0×10^7 cells contains = $1.4 \times 10^8 \ pg$ DNA

So, one cell contains = $(1.4 \times 10^8 / 6.0 \times 10^7) \ pg$ DNA

= $2.33 \ pg$ DNA

So, the amount of genomic DNA in the haploid set of chromosomes would be

$$= \frac{2.33}{2} \, pg \, DNA$$

$$= 1.17$$

$$\cong 1.2 \, pg \, DNA$$

$$= (1.2 \times 978) \, Mb$$

$$= 1173.6 \, Mb$$

Therefore, the genome size of *Etlingera assamica* is 1.2 *pg* or 1173.6 Mb.

References

Dellaporta, S.L., Wood, J. and Hicks, J.B., 1983. A plant DNA mini-preparation: Version II. *Plant Mol. Biol. Reptr.*, 1: 19–21.

Doyle, J.J. and Doyle, J.L., 1987. A rapid DNA isolation procedure from small quanities of fresh leaf tissue. *Phytochem. Bull.*, 19: 11–15.

Graham, M.J., Nickell, C.D. and Rayburn, A.L., 1994. Relationship between genome size and maturity group in soybean. *Theoretical and Applied Genetics,* 88: 429–432.

Khanuja, S.P.S., Shasany, A.K., Darokar, M.P. and Kumar, S., 1999. Rapid isolation of DNA from dry and fresh samples of plants producing large amounts of secondary metabolites and essential oils. *Plant Mol. Biol. Reptr.*, 17: 1–7.

Laurie, D.A. and Bennett, D., 1985. Nuclear DNA content in the genera *Zea* and *Sorghum.* Intergeneric, interspecific and intraspecific variation. *Heredity,* 55: 307–313.

Miksche, J.P., 1968. Quantitative study of intraspecific variation of DNA per cell in *Picea glauca* and *Pinus banksiana. Canadian Journal of Genetics and Cytology,* 10: 590–600.

Porebski, S., Bailey, L.G. and Baum, B.R., 1997. Modification of a CTAB DNA extraction protocol for plants containing high polysaccharide and polyphenols components. *Plant Mol. Biol. Reptr.*, 15: 8–15.

Rayburn, A.L., Price, H.J., Smith, J.D. and Gold, J.R., 1985. C-band heterochromatin and DNA content in *Zea mays*. *Americn Journal of Botany*, 72: 1610–1617.

Rayburn, A.L., Biradar, D.P., Bullock, D.G., Nelson, R.L., Gourmet, C. and Wetzel, J.B., 1997. Nuclear DNA content diversity in Chinese soybean introductions. *Annals of Botany*, 80: 321–325.

Turrell, F.M., 1934. Leaf surface of a twentyone-year-old catalpa tree. *Iowa Acad. Sci. Proc.*, 41: 79–84.

Ved Prakash, and Tripathi, S., 1998. Studies on Zingiberaceae of N.E. India: II. Notes on *Etlingera* Giseke. *Rheedea,* 8(2): 173–178.

Chapter 10

Some Issues Related with the Tour from Traditional Phytotherapy to Nanomedicine

Durga Prasad Gogoi

ABSTRACT

Nanotechnology is considered to be the future technology based on modern concepts and techniques. On the other hand the traditional phytotherapy is mostly based upon old traditional concepts. A basic concept of nanomedicine as a product of Nanotechnology is discussed in this paper. The aggregation morphology of molecules in Traditional Chinese medicine may be a promising way to study the mechanism of Traditional medicine, even to develop an approach of new nanomedicine. The molecular aggregation morphology of traditional medicine in the light of aggregated nanoparticles is also discussed here. In this paper, a humble attempt is undertaken to discuss few ethical issues regarding traditional medicine versus nanomedicine. This may bring innovative ideas in both the areas.

Keywords: *Nanomedicine, Aggregated nanoparticles, Chinese Traditional Medicine (CTM), ZnO nanostructure.*

Introduction

Nanotechnology is regarded as future technology. In recent years, most of the researchers have been engaged in the field of nanomedicine. Out of these, targeted drug delivery and molecular level surgery may be considered as major challenges. On the other hand Traditional phytotherapy is also is also changing its dimensions in different respects. With changing time traditional knowledge also changes.

According to R.A. Freitas nanomedicine is (1) the comprehensive monitoring, control, construction, repair, defense, and improvement of all human biological systems, working from the molecular level, using engineered nanodevices and nanostructures; (2) the science and technology of diagnosing, treating, and preventing disease and traumatic injury, of relieving pain, and of preserving and improving human health, using molecular tools and molecular knowledge of the human body; (3) the employment of molecular machine systems to address medical problems, using molecular knowledge to maintain and improve human health at the molecular scale (Freitas, 1999). According to this vision, nanomedicine will change in a short, middle, and long term perspective the life of individuals and societies in many regards. These changes are of major ethical relevance as far as they concern human self awareness, as well as the respect, responsibility, and care we owe to each other and to the environment.

A promising approach for understanding the mechanism of Traditional Chinese Medicine by the aggregation morphology was done by Juan Hu *et al.* of Pharmacy Department of Fujian University of Traditional Chinese Medicine, China. The research group has found that the Traditional Chinese Medicine (TCM) form aggregates in the aqueous solution, and the activities of two Chinese herbal formulae against three cardiovascular targets were aggregates-related. They also further studied the molecular morphology composed of aggregation and single active molecule in TCM. In the result Three kinds of aggregation processes by the bioactive molecules in the solution were elucidated: (1) the aggregation of single molecule oneself; (2) the aggregation between different single molecules; (3) the aggregation between different single molecules and the primary metabolites. Furthermore, the therapeutic activity of PUE solution was aggregates-related *in vivo* level. It was also concluded that the aggregation morphology of molecules in TCM might be a promising way to study the mechanism of TCM, even to develop an approach of new nanomedicine of TCM.

In this paper fabrication of nanoparticles through chemical reaction is highlighted and aggregation morphology of the aggregated nanoparticles is also discussed. Further approach is required to evaluate relationship among traditional phytotherapy and aggregated nanoparticles.

Traditional Phytotherapy Vs Nanomedicine

Traditional plant-derived medicines continue to offer a rich and largely unexplored source of therapeutic opportunities for the pharmaceutical and food industries. In recent years, many new health foods and herbal medicinal products have been introduced to the market. Not all are backed by appropriate research on quality, safety and therapeutic effectiveness and yet herbal remedies are a popular choice for healthcare, used by an increasing proportion of the population. The latest figures suggest 25 per cent of the UK currently uses herbal supplements while a further 25 per cent have used or would use them. New developments in recent years have seen more stringent standards brought in for such products. The MHRA recently registered the 45th product under a new scheme that requires herbal medical products to comply with quality and safety standards as they are used for medicines, in fact making such products medicine. In fact leading pharmaceutical companies can see the benefit of natural product drugs, but are reluctant to develop them as it can take decades to bring a product to market. The impact of natural product drugs and "biological tools has been increasing in importance in recent years, including major plant-derived drugs and semi-synthetic derivatives of plant compounds. According to Dr. Jen Tan, Medical Director at A.Vogel, "The journey from plants to medicines needs to be developed and regulated with properly conducted scientific research and ongoing surveillance". New research at the Centre for Pharmacognosy and Phytotherapy of UK has highlighted the importance of, for example, the composition of Echinacea preparations in determining their effect on the human body. Professor Michael Heinrich, head of the Centre, said: "This also shows the need for significant investment by the industry in assessing a products quality and safety" (School Notes, 2009).

In a study, by Je-Ruei Liu *et al.* (200) the efficacies of medicinal plant materials prepared using nanotechnology or traditional grinding were compared by examining antioxidant activity through three biological assays measuring DPPH free radical scavenging, ferrous ion chelation, and reduction power. Stronger antioxidant bioactivities were observed for the extracts prepared using nanotechnology in all tested assays. The research group has suggested that greater liberation of active components in the Danshen samples prepared using the modern technique. According to pharmacological investigation, the active constituents of Danshen can be divided into two groups: hydrophilic (*e.g.,* phenolic acids) and lipophilic (*e.g.,* tanshinones) (Kang *et al.,* 2004; Zhou *et al.,* 2006; Liu *et al.,* 2006). Various *in vitro* and *in vivo* pharmacological experiments have demonstrated that both hydrophilic and lipophilic compounds can improve microcirculation, increase the blood flow and prevent myocardial ischemia and also display antioxidant, antiatherosclerosis and antiphlegmonosis bioactivities.

Development of ZnO Based Nanomedicicine by Traditional Grinding Method

Many reports are available for the synthesis of ZnO nanoparticles. One of the easy, low cost and user friendly method is the solid state reaction method where the basic compounds are grinded in a mortar for about two hours at room temperature. Finally powdered ZnO nanoparticles (Figure 10.1B) are obtained. When enlarged image of the powder is observed though Transmission Electron Microscopy (TEM) it is found in the form of rods (Figure 10.1C).

ZnO is well known for its nontoxic, transparent, antimicrobial activity it is compatible with skin, and is used as an UV-blocker in sunscreens. Zinc Oxide has been used for many years in a wide range of cosmetic products; *e.g.,* moisturizers, lip products, foundations, make-up bases, face powders, hand creams, etc. The availability of Micronisers transparent Nano-sized Zinc Oxide has enhanced the effectiveness of Zinc Oxide and overcome the "whitening" aspect of normal Zinc Oxide powder. Consumers have the benefit of a transparent natural product with high levels of protection from UVB/UVA sunlight, making the product aesthetically pleasing. With these facts ZnO nanoparticles are widely used as biomedicine.

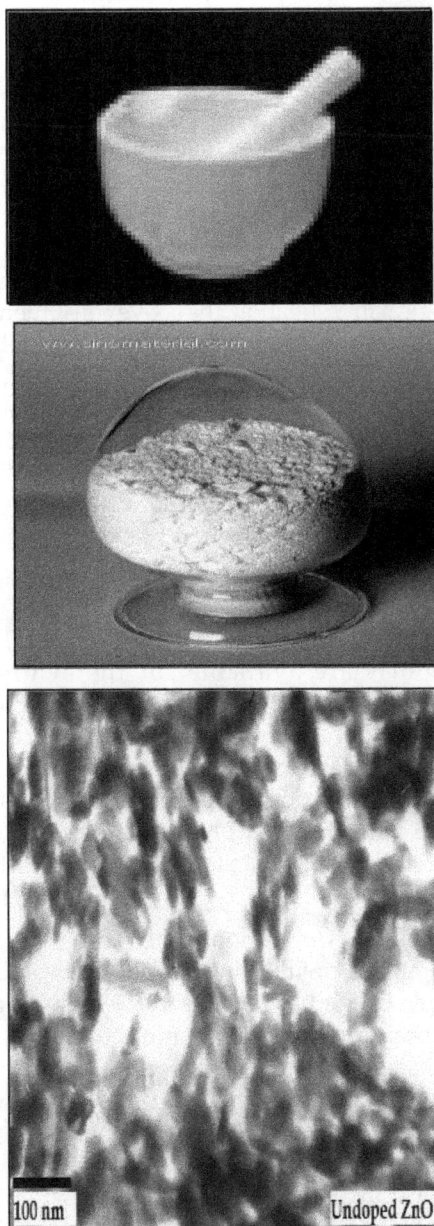

**Figure 10.1: A: Mortar Grinder, B: ZnO Nanopowder,
C: ZnO Nanorods (TEM)**

Conclusion

Some issues regarding Traditional Phytotherapy and nanomedicine have been discussed in the paper. Traditional process is easier than modern methods of nanomedicine but due to the lack of systematic synthesis and scientific evaluation required target may not be achieved. Study reveals that in some of the synthesis process of modern nanomedicine the same traditional grinding method for preparation of nanoparticles has to be followed. Scientific systematic evaluation is required to develop Traditional Phytotherapy. This area of research may definitely bring drastic change in the field of traditional phytotherapy.

References

Chun and Li, L.N., 1988. *J. Nat. Prod.,* 51: 145–149.

Freitas, R.A., 1999. *Nanomedicine, Vol. 1: Basic Capabilities.* Georgetown, Texas, p. 418.

Liu, Je-Ruei, Chen, Guo-Feng, Shih, Hui-Nung and Kuo, Ping-Chung, 2008. *Phytomedicine,*15: 1–2: 23–30.

School Notes, October 2009, page–2 published by The School of Pharmacy, University of London.

Chapter 11

A Correlation Between Fluorescence of Chlorophyll Solution in Some Medicinal Plant Leaves and Absorption in Liquid Phase

Mitali Konwar

ABSTRACT

Red chlorophyll fluorescence of a number of medicinal plants Azardicta indica, Clerodendron colebrookonium, Leucas linifolia, Mesua ferua, Leucas cephalotes/lini folia have been photographed with the help of a two prism glass spectrograph and a 500W hellogen lamp as continuum. The corresponding absorption spectra of this sample have also been photographed. The salient points of the observation are that the fluorescence intensity depends on the concentration of the sample but at the same time it also depends on the absorption of the sample. When the red fluorescence intensity is maxima the absorption band also exhibits maximum extinction. The fluorescence intensity has been estimated with the help of a photodiode connected to a optical fiber. The result so obtained in the work can be used to estimate the fluorescence content of the solution of the medicinal plants. We have also used Ar^+ laser to excite fluorescence spectra and worked out a correlation between absorption and fluorescence.

Introduction

Fluorescence is a process in which an atom or molecule emits radiation in the course of a transition from a higher to a lower electronic state. A more restricted definition, applicable particularly to atomic processes, excludes the special case, known as resonance radiation, in which the wavelength of the emitted radiation is the same as that of the exciting radiation. The term fluorescence is further restricted to phenomena in which the time interval between the acts of excitation and emission is small, of excitation and emission is small, of the order of 10^{-8}–0^{-3} second. This distinguishes fluorescence from phosphorescence, where the time interval between absorption and emission may extend from 10^{-3} second to several hours. The phenomenon of fluorescence was known by the middle of the century. It was the British scientist Stokes who first made the observation that the fluorescing light has longer wavelengths than the excitation light, a phenomenon that has become to be known as Stokes-shift. Fluorescence microscopy is an excellent method of studying material that can be made to fluoresce either in its natural form or when treated with chemicals capable of fluorescing. One of the most prominent processes is the chlorophyll fluorescence, which is readily observed when a solution of chlorophyll (green leaves) is illuminated with a strong source of continuum radiation. The red colour characteristics of fluorescence readily appear in a sample cuvette. Light energy is absorbed by chlorophyll within plant tissues and used to drive photochemistry of photosynthesis and thus become chemical energy available to the plant for growth. Light in the waveband 400-700 nm is absorbed by chlorophyll and used for photochemistry. This light is termed photosynthetic ally Active Radiation (PAR). Although fluorescence emission from whole leaf system is too weak to be viewed with the naked eye, it can be observed from the illuminated extracts of a chlorophyll solution. Peak fluorescence occurs in the red region of the spectrum (685 nm) and extends into the infrared region to around 800nm. The fluorescence from chlorophyll has been used extensively in the past to characterize and investigate agricultural plants [1–5]. In the present work we report an experimental investigation, which correlates between absorption and fluorescence for the same chlorophyll extracts of medicinal plants. This correlation is not exactly known in the investigations of the early workers. The approach presented in this work is simple, as we have used the linear dimension of the absorption and the intensity of the fluorescence for our correlation.

Experimental

We have used five specimens of medicinal plants namely (a) *Azardicta Indica*, (b) *Clerodendron colebrookorium*, (c) *Leuca linifolia*, (d) *Mesua ferua L* and (e) *Leucas cephalotes-linifolia* for our studies of fluorescence and the corresponding absorption. In all the cases immersion of the crushed leaves in a glass vessel containing a suitable solvent like acetone enables the pigments responsible for their green colour to be conveniently and quickly extracted. After filtering, the extract may be transferred to a cuvette of suitable length. The cuvette with the solution is held in front of the slit of a two prism glass spectrograph. A bright source of halogen lamp (500W) is used as the source of continuum radiation for observing fluorescence and as well as absorption. The absorption spectrum of a sample solution may be visually observed on the position of the plate holder with the help of a ground glass.

Figure 11.1 shows the characteristic picture of the red coloured fluorescence, which appears when the radiation is incident on it. This is the well-known chlorophyll fluorescence. It is worthwhile to observe that the red colour becomes very intense when the concentration of the extract is more and at the same time the

Figure 11.1: Well Known Red Fluorescence from a Chlorophyll Extract

corresponding absorption is maximum. Figure 11.2 shows the absorption and fluorescence spectra of five specimens when the concentration of the extract is maximum in each case. For our correlation between absorption and fluorescence we have used a graphical method, which relates the linear dimension of the absorption band and fluorescence intensity. It is observed that (and it is usually the case) absorption length is minimum when the concentration is very low. The corresponding fluorescence intensity is also very low. The intensity of the fluorescence is measured with the help of a photodiode connected through an optical fiber. Figure 11.3 shows the correlation between the fluorescence intensity and absorption lengths. The specimen used in this case is *clerodendron colebrookorium*. Similar graphs may also be obtained for other samples.

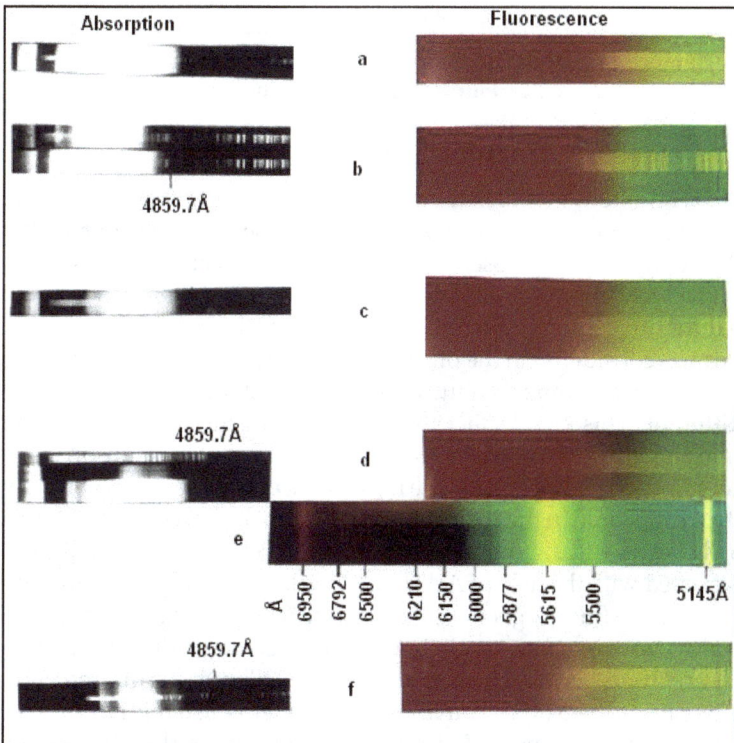

Figure 11.2: Absorption and the Corresponding Fluorescence from Five Chlorophyll Extracts of Medicinal Plants

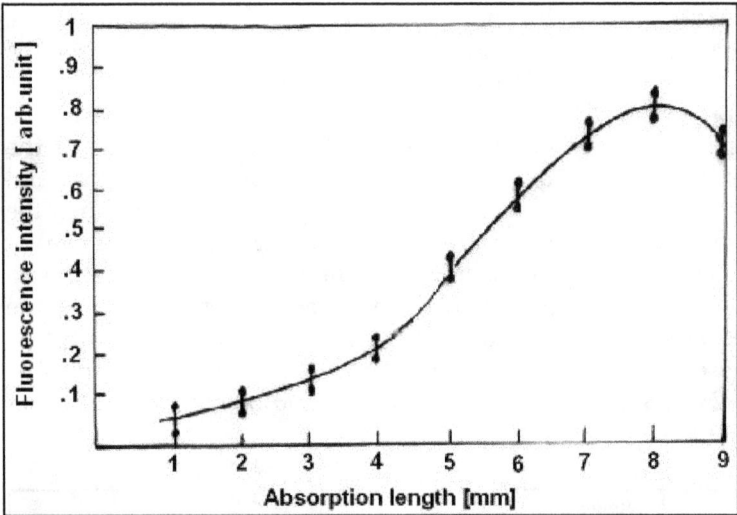

Figure 11.3: Relation Between Absorption Bands Width and Fluorescence Intensity

Results and Discussion

The extent to which the red fluorescence colour penetrates in the cell before the green colour appears depends on the particular chlorophyll extract chosen. As for example for a cell of 15mm breath the red and green colour combination in terms of their penetration depths are shown in the following Table 11.1 for five chlorophyll extracts considered in the present work. It is scarcely surprising that the red green combinations for different chlorophyll extracts are different. This simple fact may be used to characterize different chlorophyll extracts by a method entirely different from absorption. As shown in Figure 11.2, all the fluorescence show nearly continuum structure while in the case of Figure 11.2 (e) fluorescence spectrum of *Mesua ferra* leaf shows more than ten discrete bands when the leaf is excited with the help of Ar⁺ laser (500w). This clearly differentiates between the fluorescence excited with the help of a broadband source and narrow band or single line source. As may be inferred from Figure 11.3 when the concentration is at maximum the fluorescence intensity as measured from the digital meter is also minimum. The Fluorescence intensity **I** gradually increases as the absorption length **L** (that is the horizontal spread of the absorption region in the spectrogram) also increases. It may be noted from the **I-L** curve that the fluorescence intensity **I**, at a particular stage, does not increase even after the increase of absorption length **L** but in reality the fluorescence intensity decreases. This occurs primarily due to self-absorption, which is always present whenever a beam of light penetrates a certain distance of the medium. As concentration of the medium is maximum the process of self-absorption is complete and under this circumstance no transmitted light is observed. We must also consider the process of scattering and fluorescence because in all the cases both may be operating. Fluorescence and absorption are familiar phenomena in nature, but surprisingly enough, attention does not appear to have been drawn to particular type of relation, which we have worked out using simple parameters.

Table 11.1: Penetration Length Versus Colour of Fluorescence Observed in a Sample Cell of Dimension 1 cm x 1 cm x 6 cm

Sl.No.	Sample	Colour	Penetration Depth
1.	*Azardicata indica*	Red	5mm
		Green	10mm
2.	*Cherodendron colebrookorium*	Red	5mm
		Green	10mm
3.	*Leucas linifolia*	Red	2mm
		Green	13mm
4.	*Mesua ferea*	Red	8mm
		Green	7mm
5.	*Leucas linifolia*	Red	11mm
		Green	4mm

Conclusion

The present experimental investigation shows that the fluorescence intensity from a chlorophyll extract may be correlated with the absorption bandwidth by a simple empirical relation, which is not linear and closer to Beer's law.

References

Buller, W.L. and Kitajima, 1975. *Biochem. Biophys. Acta.*, 396: 72–85.

Chappelle, E.W., McMurtrey III, J.E. and Kim, M.S., 1991. *Remote Sens. Environ.*, 36: 213–218.

Krause, G.H. and Weis, E., 1991. *Ann. Rev. Plant Physiol. Plant Mol. Biol.*, 42: 313–349.

Rosema, A. and Zahn, H., 1997. *Remote Sen. Environ.*, 62: 101–108.

Weis, E. and Berry, J.A., 1940. *Biochem. Biophys. Acta.*, 894: 198–208.

Chapter 12

Indigenous Knowledge System in Traditional Phytotherapy of North East India: Priority Issues

R. Kandali, R.K. Goswami,
B.K. Konwar and J. Buragohain

ABSTRACT

The North eastern region of India is well known for its medicinal plant diversity and the age-old healthcare traditions. The need of the hour is to establish these traditional values in both national and international perspectives realizing the ongoing developmental trends in traditional knowledge. The most important advantage of herbal medicine is the minimal side effect and relatively low cost as compared to synthetic medicines. People have begun to realize the benefits associated with natural remedies again. The continuous increase in the demand and expanding trade on medicinal and aromatic plants worldwide, have jeopardized the survival of several plant species due to indiscriminate harvesting of natural flora including those in forests. A sustainable growth in this sector will improve the socio-economic condition of the people of the locality through self employment and entrepreneurship development, besides helping conservation of traditional knowledge base and culture of the indigenous tribal communities in the region. The issue of IPR with respect to ethno-medicinal bio resources has received considerable attention in the recent past. Trade Related Aspects of Intellectual Property Rights (TRIPS) agreement is one of the important agreement of World Trade Organization (WTO) and it specifies the standard and requirements of the IPR. This paper addresses the issues related to conservation of medicinal plant diversity along with traditional knowledge base.

Introduction

The plant kingdom contributed immensely to human health from the time immemorial when no synthetic medicines were available, and when no concept of surgery existed. There is a growing demand today for plant-based medicines, health products, pharmaceuticals, food supplements etc globally. The North Eastern Region of India forms a distinctive part of the Indo-Burma hotspot which ranks the sixth among the 25 biodiversity hotspots of the world and is the prime one among the two identified in the Indian subcontinent. Since, north eastern region of India is well known for its medicinal plant diversity and the age-old healthcare traditions, there is an urgent need to establish these traditional values in both national and international perspectives realizing the ongoing developmental trends in traditional knowledge. Traditional knowledge on medicinal plants and their use by indigenous cultures are not only useful in the conservation of traditions and biodiversity but also for the community healthcare and drug development in the present and future. Apart from the tribal groups, many other forest dwellers and rural people also possess unique knowledge about herbal therapy, particularly on locally available plant species. Due to lack of interest among the younger generation as well as their tendency to migrate to cities for lucrative jobs, this great wealth of knowledge is declining. Right from its beginning, the documentation of traditional knowledge especially on the medicinal uses of plants, has provided many important drugs of modern day. But, information on the use of plants for medicine and molecular parameters present in them from this area of the country is rare. Thus, many important leads to drug discovery may be lost in the absence of proper documentation and identification of the medicinal plants.

Appraisal of the Knowledge Base

The North Eastern Region of India harbours more than 130 tribal communities. In general, the tribes of North East India have been categorized into two broad groups: Khasi and Jaintia tribes of Meghalaya, who belong to 'Monkhemar' culture of Austric dialect and the rest of the tribal groups are basically Mongoloid, who belongs to Tibeto-Burman subfamily of Tibeto-Chinese group. Sajem and Gosai (2006) studied the traditional use of medicinal plants by the Jaintia tribes in North Cachar Hills district of Assam. Altogether, 30 types of ailments have been reported to be cured by using 39 medicinal plant species. For curing diverse form of ailments, the use of aboveground plant parts was higher (76.59 per cent) than the underground parts (23.41 per cent). Of the aboveground parts, leaf was used in majority of cases (23 species), followed by fruit. Different underground parts such as root, tuber, rhizome, bulb and pseudo-bulb were also found to be in use by the Jaintia tribe as medicine. Begum and Nath (2000) studied the medicinal plants used for skin diseases and related problems in North Eastern India. According to them, out of 275 plant species, 224 are used for treatment of specific human ailments such as allergies, burns, cuts and wounds, inflammation, leprosy, leucoderma, scabies, smallpox and sexually transmitted diseases. Some of the plant species, including *Artemisia nilagirica* (CI) Pamp., *Calotropis gigantea* (L) R. Br., *Cannabis sativa* L., *Cassia alata* L., *C. fistula* L., *Centella asiatica* L., *Cyclea peltata* Hk., *Datura metal* L., *Drymaria cordata* (L.) Willd. ex Roam and Schult., *Jatropha curcus* L., *Litsea cubeba* Pers., *Mimosa pudica* L., *Plantago major* L. and *Plumeria acutifolia* Ait. are used for disease treatments by different ethnic groups. Temjenmongla and Yadav (2005) studied the anticestodal efficacy of folklore-based medicinal plants of Naga tribes in North East India. They found that the leaves of *Psidium guajava*, *Houttuynia cordata* and stalk of *Lasia spinosa* possess a profound anticestodal efficacy as evident by the mean mortality time of *R. echinobothrida*. Moderate activity was recorded for the leaves of *Clerodendrum colebrookianum, Lasia spinosa* and *Centella asiatica*, while *Curcuma longa, Cinnamomum cassia, Gynura angulosa, Lasia spinosa* (stem) and *Aloe vera* revealed a negligible degree of anticestodal activity. Das *et al.* (2006) studied the medicinal plants of North Kamrup district of Assam used in primary healthcare system. They found that out of 31 medicinal plants, 8 are used in stomach disorder, 4 in body pain, 3 in piles, 2 in skin disease, 2 in ulcer and remaining in dysurea, boils, nervous affection, spermatorrhoea, jaundice, toothache, hydrophobia, sinusitis, asthmatic trouble and obstetrics problem. Borah *et al.* (2006) studied the traditional medicine in the treatment of gastrointestinal diseases in the Upper Assam. The results revealed use of 38 plant species represented by 36 genera and 29 families for the treatment of various gastrointestinal diseases. Das *et al.* (2005) reported the use of *Jatropha curcus* Lin. for the treatment of dysmenorrhoea by the Koch-Rajbongshi tribe in Nalbari district of Assam. Borah *et al.* (2001) highlighted the use of 25 medicinal plants by the rural and tribal communities of Darrang district of Assam. They reported that *Spondias pinnata* (syn. *S. mangifera*) fruit and bark was used to cure dysentery. Jalil *et al.* (2006) reported 38 medicinal plant species have been used as medicine for several common disease like bronchitis, asthma, hemorroids, dysentery and cardiac disorder, etc. by the adivasi (saontal) people of Dhubri district of Assam. Choudhury *et al.* (2006) reported analgesic activity of *Dioscorea hamiltoni* Hook., *Ipomea aquatica* L., *Musa paradisiacal* L., *Solanum indicum* L. and *Solanum torvum* Swartz., medicinal plants of Tripura. The state of Nagaland is inhabited by 16 major distinct aboriginal Naga tribes. 85 per cent of these tribes live in rural areas. They use the plant *Blechnum orientale* for urinary bladder problem and as germicide. They also use *Taxus Bacccata* to treat tumours and chronic diseases Jamir (2006). Tai Khamti is an important tribe of Arunachal Pradesh which has a practice of rich traditional biodiversity conservation specially plants like *Ficus* spp. among themselves (Gogoi, 2006). The Bawm tribal community of Mizoram utilizes 21 plant species for herbal medicine. They have their own system of preservation and utilization of their valuable bio-resources (Lalnunmawia *et al.,* 2006).

People Awareness

There is a global resurgence of interest in the alternative medicinal health care systems through traditional herbs during last couple of decades. Many advantages of such eco-friendly traditions exist. The plants used for various therapies are readily available, easy to transport and have a relatively long shelf life. The most important advantage of herbal medicine is the minimal side effect and relatively low cost as compared to synthetic medicines. People have begun to realize the benefits associated with natural remedies again. Medicinal plants work in an *integrated* or *pro-biotic approach* and with little adverse effects. For example, a regular intake of garlic can control cholesterol and high blood pressure within a moderate period of time, but taking synthetic drugs make the person's body completely dependent on that particular medicine (Babu and Madhavi, 2001). Allopathic medicines may cure a wide range of diseases; however, its high prices and side effects are causing many people to return to herbal medicines.

Conservation Issues

The population of several plant species has been reduced and in fact some of them are being extinct due to indiscriminate harvesting of natural flora including those in forests. It was estimated that about 90 per cent collection of medicinal plants is from wild source and 70 per cent of collections involve destructive harvesting. As a result of that, many species become extinct and some are endangered. The International Union for the Conservation of Nature and Natural Resources (IUCN) is the world's largest and most important conservation network. The Union brings together 82 States, 111 government agencies, more than 800 non-governmental organizations (NGOs), and some 10,000 scientists and experts from 181 countries in a unique worldwide partnership. The Union's mission is to influence, encourage, and assist societies throughout the world to conserve the integrity and diversity of nature and to ensure that any use of natural resources is equitable and ecologically sustainable. The **Medicinal Plant Specialist Group** (MPSG) is a global network of experts contributing within their own institutions and in their own regions to the conservation and sustainable use of medicinal plants. The MPSG was founded in 1994, under the auspices of the Species Survival Commission (SSC) of the IUCN, to increase global awareness of conservation threats to medicinal plants, and to promote sustainable use and conservation action. TRAFFIC, the wildlife trade monitoring network, works to ensure that trade in wild plants and animals is not a threat to the conservation of nature. TRAFFIC is a joint programme of WWF and IUCN - The World Conservation Union. The Medicinal Plant Specialist Group works with TRAFFIC on medicinal plants in trade, contributing to TRAFFIC's vision of a world in which trade in wild plants and animals will be managed at sustainable levels without damaging the integrity of ecological systems and in such a manner that it makes a significant contribution to human needs, supports local and national economies and helps to motivate commitments to the conservation of wild species and their habitats. Formed as a series of measures at a meeting of the World Conservation Union (IUCN) in Washington DC in 1963, the CITES (Convention on International Trade in Endangered Species of Wild Fauna and Flora) treaty is an inter-governmental agreement which exists to protect plants and animals from over-exploitation and serious decline via the actions of unscrupulous traders and rogue elements. CITES is now supported by 189 nations, and offers a complex legal regulatory structure from which it seeks to impose its effects.

Barbhuiya *et al.* (2009) prioritized 24 medeicinal plant species of Barak Valley of Assam for conservation. Among these, the population of *Acorus calamus* Linn., *Aegle marmelos* Linn., *Rauvolfia serpentina* (L.) Benth ex. Kurz are reducing day by day for over exploitation of medicinal purposes, felling of trees for timbers and poor inherent regeneration ability. They further stressed the need for comprehensive herbal policy for public awareness, cultivation and conservation on a sustainable basis within the environmentally protected regime. Kala *et al.* (2006) reported that information on the propagation of medicinal plants is available for less than 10 per cent and agro-technology is available only for 1 per cent of the total known plants globally. This trend shows that developing agro-technology should be one of the thrust areas for research. Furthermore, in order to meet the escalating demand of medicinal plants, farming of these plant species is imperative. Apart from meeting the present demand, farming may conserve the wild genetic diversity of medicinal plants. Farming permits the production of uniform material, from which standardized products can be consistently obtained. Cultivation also permits better species identification, improved quality control, and increased prospects for genetic improvements. Selection of planting material for large-scale farming is also an important task. The planting material therefore should be of good quality, rich in active ingredients, pest- and disease-resistant and environmental tolerant. For the large scale farming, one has to find out whether monoculture is the right way to cultivate all medicinal plants or one has to promote polyculture model for better production of medicinal plants.

Certain anthropogenic activities such as deforestation, shifting cultivation, etc. were the main causes to affect the medicinal plant diversity in Assam (Das *et. al.,* 2008). They further urged to adopt appropriate measures to improve the habitat of these wild medicinal plants by controlling deforestation, soil erosion, etc. They advocated sustainable harvesting of medicinal plants, establishment of botanical gardens and encouragement for *in vitro* conservation of rare and endangered medicinal plant species of Assam through modern breeding techniques. In the premier institutes of this region works are conducted for propagation of important medicinal plants, *i.e., Andrographis paniculata, Hygrochilus parishii* through tissue culture technique (Purukayastha *et al.,* 2006).

Towards Sustainable Use

For the sustainable growth in this sector, some necessary steps have to be initiated. Firstly, necessary legislative, legal and administrative legislations have to be made to prohibit the collection of those wild plant species which are rare, threatened, endangered and vulnerable. Secondly, the indigenous knowledge system has to be well documented. The traditional medicinal (herbal) practitioners should be given due recognition, their efforts and activities supported by helping them to tie up with research institutions. Thirdly, biochemistry and biotechnology of wild economic plants be pursued for future need-base commercial access and a chain of botanical gardens be established at different altitudes to conserve and multiply the germplasm of rare and endangered species.

Legal Issues

The legal rights accrued on intellectual property are termed as intellectual property rights (IPR) which is a legal protection for intellectual property given by the state. The issue of IPR with respect to ethno-medicinal bio resources has received considerable attention in the recent past. Trade Related Aspects of Intellectual Property Rights (TRIPS) agreement is one of the important agreement of World Trade Organization (WTO) and it specifies the standard and requirements of the IPR. The article 27.3 (b) is worth recalling as it brings the bio-resources including plants and animals other than microorganisms under the preview of IPR. Under this article, member countries shall provide patents or effective *sui generis* system or combination of both for plant variety protection. The U N convention on biological diversity also contains provision for integrating World Trade, IPR and biodiversity. Traditional knowledge associated with biological resources is an inseparable component of the resource itself. The traditional knowledge that is developed by each ethnic tribe can provide important lead towards development of a useful product or process. During this process, out of the total commercial profits obtained by the multinational companies a sizeable share should go to the preservers of biodiversity and practitioners of traditional knowledge. The Biodiversity Act 2002 framed many rules for the sustainable utilization of medicinal plants (Puspangadan, 2005).

Conclusion

There is an urgent need for comprehensive herbal policy for public awareness, conservation of medicinal plants on a sustainable basis within the environmentally protected regime and to provide legal protection to the existing traditional knowledge base. There is a tremendous scope for enhancing the production and export of these medicinal plants and their products which will help to earn foreign exchange. A sustainable growth in this sector will improve the socio-economic condition of the people of the locality through self employment and entrepreneurship development, besides helping conservation of traditional knowledge base and culture of the indigenous tribal communities in the region.

References

Babu, S.S. and Madhavi, M., 2001. In: *Medicinal Plants and Herbs: Green Remedies.* Pustak Mahal, New Delhi, India.

Barbhuiya, A.R., Sharma, G.D., Arunachalam, A. and Deb, S., 2009. Diversity and conservation of medicinal plants in Barak Valley, Northeast India. *Indian J. Traditional Knowledge*, 8(2): 169–175.

Begum, D. and Nath, S.B., 2000. Ethnobotanical review of medicinal plants used for skin diseases and related problems in North Eastern India. *Journal of Herbs, Spices and Medicinal Plants*, 7(3): 55–93.

Borah, P.K., Gogoi, P., Phukan, A.C. and Mahanta, J., 2006. Traditional medicine in the treatment of gastrointestinal diseases in Upper Assam. *Indian Journal of Traditional Knowledge*, 5(4): 510–512.

Choudhury, R., Choudhury, M.D. and Paul, S.B., 2006. Analgesic activity of certain ethnobotanical plants of Tripura state, India.

Abstract of paper presented in *National Seminar on Value Addition to Bioresources of N.E. India: Postharvest Technology and Cold Chain,* Organized by Department of Botany, Gauhati University and ASTEC, Govt. of Assam during 19–21 May, pp. 122.

Das, A.K., Dutta, B.K. and Sarma, G.D., 2008. Medicinal plants used by different tribes of Kachar district, Assam. *Indian J. Traditional Knowledge,* 7(3): 446–454.

Das, N.J.S., Saikia, P., Sarkar, S. and Devi, K., 2006. Medicinal plants of North-Kamrup district of Assam used in primary healthcare system. *Indian Journal of Traditional Knowledge,* 5(4): 489–493.

Gogoi, P., 2006. Traditional biodiversity conservation among the Tai Khamti tribes of Arunachal Pradesh. Abstract of paper presented in *National Seminar on Biodiversity and Indigenous Knowledge System,* sponsored by UGC held during October, 25–26 at RGU, Itanagar, Arunachal Pradesh, India, pp. 10.

Jamir, N.S., 2006. Ethno-biodiversity and its conservation in the state of Nagaland. Abstract of paper presented in *National Seminar on Biodiversity and Indigenous Knowledge System,* sponsored by UGC held during October, 25–26 at RGU, Itanagar, Arunachal Pradesh, India, pp. 23.

Jalil, A., Devi, N., Sarma, G.C. and Rahman, A., 2006. Study of some common medicinal plants used by Adivasi (Saontal) people of Dhubri district of Assam. Abstract of paper presented in *National Seminar on Value Addition to Bioresources of N.E. India: Postharvest Technology and Cold Chain,* Organized by Department of Botany, Gauhati University and ASTEC, Govt. of Assam during 19–21 May, pp. 122.

Kala, C.P., Dhyani, P.P. and Sajwan, B.S., 2006. Developing the medicinal plants sector in Nothern India: Challenges and opportunities. *J. Ethnobiology and Ethno-medicine,* 32(2).

Lalnunmawia, F., Lalfakzuala, R. and Monlai, S., 2006. Indigenous knowledge system and sustainable utilization of bio-resources. Abstract of paper presented in *National Seminar on Biodiversity and Indigenous Knowledge System,* sponsored by UGC held during October, 25–26 at RGU, Itanagar, Arunachal Pradesh, India, pp. 14.

Purukayastha, J., Sugla, T., Bakshi, S., Solleti, S., and Sahoo, L., 2006. Rapid *in vitro* plant regeneration from nodal explant of *Andrographis paniculata* Nees: A valuable medicinal plant. Abstract of paper presented in *National Seminar on Biodiversity and Indigenous Knowledge System,* sponsored by UGC held during October, 25–26 at RGU, Itanagar, Arunachal Pradesh, India, pp. 16.

Puspangadan, P. and Kumar, B., 2005. Ethnobotany, CBD, WTO and the Biodiversity Act of India. *Ethnobotany,* 17: 2–12.

Sajem, A.L. and Gosai, K., 2006. Traditional use of medicinal plants by the Jaintia tribes in North Cachar Hills district of Assam, North East India. *Journal of Ethnobiology and Ethno-medicine,* 2: 33.

Temjenmongla and Yadav, A.K., 2005. Anticestodal efficacy of folklore medicinal plants of naga tribes in north-east India. *Afr. J. Traditional Complimentary and Alternative Medicine,* 2(2): 129–133.

Chapter 13

Designing a Web-Enabled Database for Ethno-medicinal Plants of Upper Brahmaputra Valley of Assam

Ridip Hazarika and Bijay Neog

ABSTRACT

Traditional medicine has always been played a key role in curing various ailments in different ethnic groups living in remote areas of Assam. This paper presents the experiences of creating a web-enabled and user-friendly information system on certain medicinal plants widely used by the traditional practitioners in the upper Brahmaputra valley of Assam. The data include geographic and temporal distributions studies along-with botanical characteristics. A brief description of various plant parts used and bioactive compounds are also been mentioned. Based on certain efficient and user friendly software with quick retrieval system, the present designs of the DBMS have been achieved. This information system would be much useful to the environmentalists, researchers, planners, administrators, decision makers and local communities to

* Corresponding author: E-mail: ridiph@rediffmail.com

understand, to comprehend existing biogenetic resources and related knowledge. Also to adapt new management practices in the changing environment for sustainable management of nature's unique genetic heritage.

Keywords: DBMS, Ethno-medicine, Upper Assam.

Introduction

North East India comprises of seven states commonly known as the "Seven Sisters". They are Arunachal Pradesh, Assam, Manipur, Meghalaya, Mizoram, Nagaland and Tripura. North East India is one of the richest biodiversity regions in India. Assam is the gateway to the north-east; it is the second largest state of north east India situated between 24°2'-27°6'N latitude and 89°8'-96°E longitude covering an area of 78,438 sq km of which 23,688 sq km area is covered by forest. There are large amount of medicinal plants specimens present in this area. The information that where specific species of plants lived, how they lived, and how they responded to changes in the environment and traditional medicinal uses are very much important for proper scientific study of these specimens. Bioactive compounds, such as those derived from plant foods, are of growing interest to the scientific community and food industry because of their putative health-promoting properties (Black *et al.,* 2008). This vast global archive of collection data is a priceless source of information on the Earth's bio-systems; however, collections of data cataloged without the aid of computers are incompatible. Bioinformatics plays an essential role in today's plant science. As the amount of data grows exponentially, there is a parallel growth in the demand for tools and methods in data management, visualization, integration, analysis, modeling, and prediction (Rhee *et al.* 2006). Information technology is used to create web-based information system, which stores, organizes and provides combined text and multimedia data on botanical, toxicological, biochemical as well as pharmacological properties of plants.

Dibrugarh University, a premier university in the northeast India has taken up such an initiative with the collaboration of the University Grant Commission, New Delhi, to prepare a digital data book of the traditionally used medicinal plants of this region. The task is aimed to digitalize the compiled information on traditionally used potential medicinal flora of upper Assam along with their geographic and temporal distributions. This web-enabled information system will assess and analyze the huge database and report the findings; that could be accessed efficiently by broad international user community. The programme would also have linkages among similar databases for the efficient retrieval of all available information about these species.

The database has been designed and made up of several efficient data capture soft wares, which would offer a great opportunity to support the management effort of inventory, monitoring conservation, rehabilitation of invaluable flora and their related issues. The initiative would be expected to promote the efficiency of exploitation and utilization of bioresearches using modern biotechnological tools. This model system will also strengthen the academic and research capability in key areas relevant to biodiversity conservation, sustainable natural resource management and restoration of degraded ecosystems and natural communities. Once developed, the mapped and tabular system information database can be shared easily on an interagency basis by the heterogeneous group of workers to serve the purpose of future policymaking on biodiversity and related issues.

Materials and Methods

In the scope mentioned above, a model was designed on Medicinal plant database system relying on object-oriented database (Olivier and Solms, 1994) technology. The system is not only restricted to special scientific users and researchers but also open to home users who interested in medical plants. Therefore, the access to this database is free of charge and the content of the database is designed in such a way that scientific users and researchers can benefit, as well as the interested home users.

WampServer (http://www.wampserver.com/en/) is used to create web enabled database system which is a combination of Apache with PHP server-side language and the MySQL database. Other languages like HTML (Hyper Text Markup Language), Java Script also used for designing purposes.. Update functionality is provided by a separate tool, available to the administrator only, which used to upload new data and images.

Results and Discussion

Following is the simplified organizational table of Medicinal Plant Database.

**Table 13.1: Simplified Organizational Table
of the Medicinal Plant Database**

Sl.No.	Primary Entity	Comment
1.	Home	About digital data book :advantages and uses
2.	About Medicinal Plant	Brief botanical description, phytochemistry, traditional medicinal uses of plants.
3.	Area	Assam with special reference to Upper Assam and Map
	Administrative page	To update information time to time
5.	About us	DU team in UGC project
6.	Photo gallery	Photos of interesting features of area
7.	Links	Related active websites
8.	Feed back/contact	Feed back format

Home page of the Medicinal Plant Database system contains all the primary entity like Home, About Medicinal Plant, Area, Photo gallery, About Us, Feedback, Administrative Page and Links. To get the information about the medicinal pants just place the mouse point on About Medicinal Plant, then two sub menus will come, *i.e.* Search by Scientific Name and Search by Local Name. Click on Search by Scientific Name, plants are alphabetically arranged on the basis of scientific name and if Search by Local Name then alphabetically arranged on the basis of local name. Clicking on the name of the plant user will get the information about brief taxonomic description, geographic distribution, phytochemistry, traditional medicinal uses, images of different plant parts and geographical position of plants. In the same way one can get information about area which contains brief description about geography, climate, population, major crops and different communities of five district of upper Brahmaputra valley, Assam namely *Tinsukia, Dibrugarh, Sivasagar, Jorhat, Golaghat.* Other entity like photo gallery contains Photos of interesting features of five districts. Links is another interesting feature of the database system through which user can go to other related active sites. In Feed back page user can give their queries and comments, which are

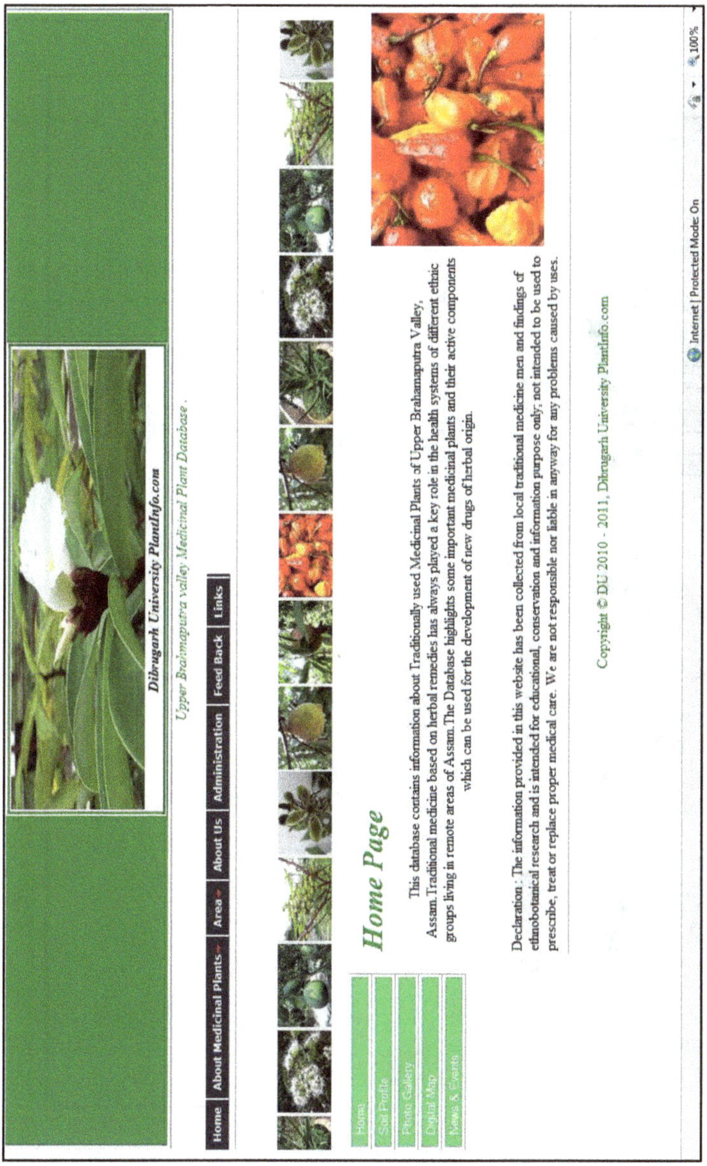

Figure 13.1: Home Page of the Medicinal Plant Database

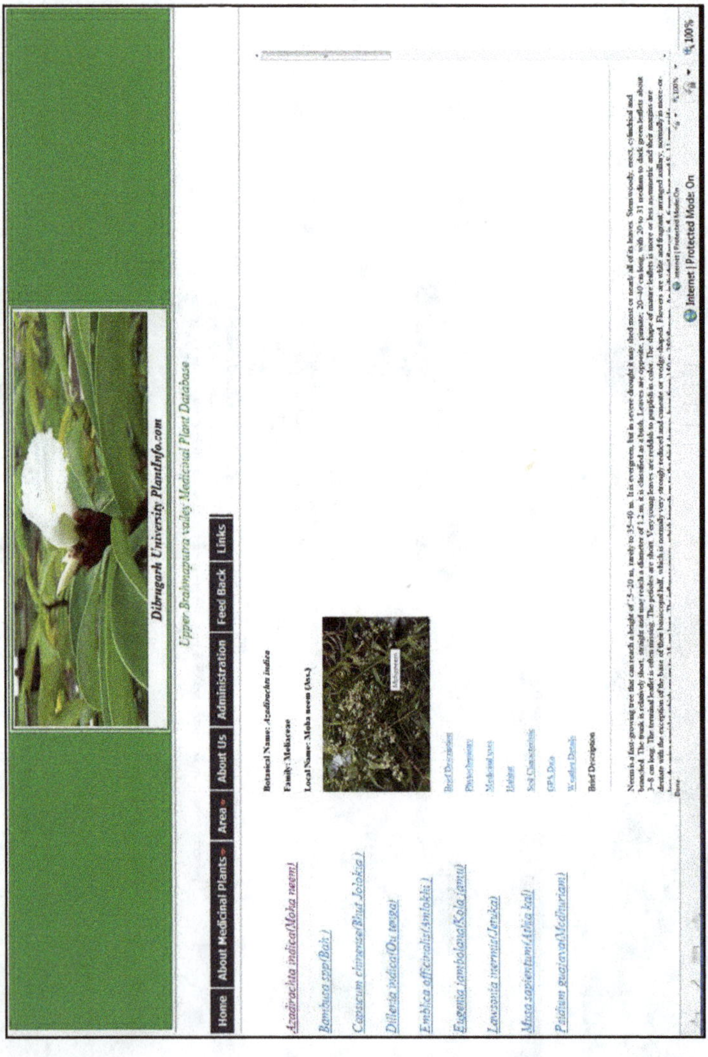

Figure 13.2: Search by Scientific Name of the Medicinal Plant

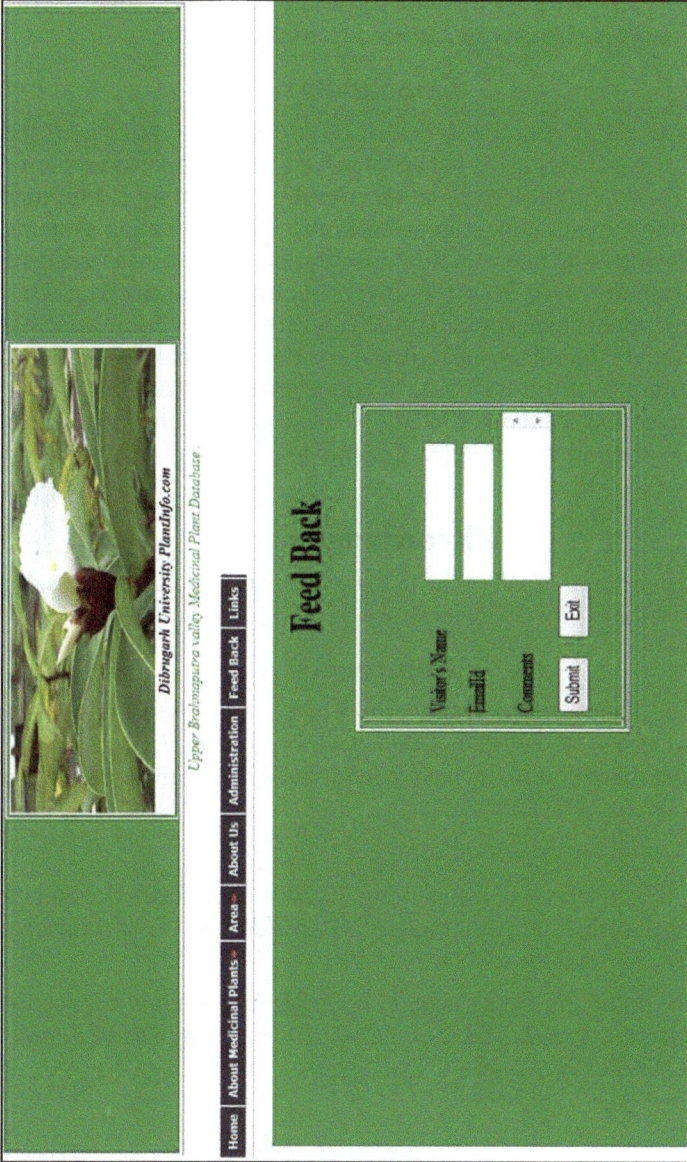

Figure 13.3: Feed Back Page

automatically store in the database after submission. Administrator page requires password to open, will be available to administrator only which used to upload updated data and images.This information will easily be accessed by touch of a button.

Conclusion

The rapid disappearance of biodiversity is causing major concern. If everything can not be conserved, prioritization is necessary for conserving at least some therapeutically potential spp. Computer aided documentation, storage and retrieval information systems could contribute towards the conservation and bioprospecting of traditionally used lesser known species of this region. This information system would be much useful to the environmentalists, researchers, planners, administrators, decision makers and local communities to understand, to comprehend existing biogenetic resources and related knowledge. These types of information also help to screen these important plants to conserve for sustainable uses for the future generation.

Acknowledgement

The authors are thankful to the University Grants Commission, New Delhi, for the financial support.

References

Black, L., Kiely, M., Kroon, P., Plumb, J. and Gry, J., 2008. Development of EuroFIR–BASIS: A composition and biological effects database for plant based bioactive compounds. *Nutrition Bulletin,* 33 (1): 58–61.

http://www.wampserver.com/en/

Rhee, S.Y., Dickerson, J. and Dong, X., 2006. Bioinformatics and its applications in plant biology. *Annual Review of Plant Biology,* 57: 335–360.

Olivier, M.S. and Sebastiaan, H.S., 1994. A taxonomy for secure object–oriented databases. *ACM Transactions on Database Systems (TODS),* 19(1): 3–46.

Chapter 14

Traditional Phytomedicine Used for Hepatitis Treatment Among the Tea Tribes Community of Upper Assam

Firoza Khatun

ABSTRACT

India is a repository of herbal medicines and there are evidences of herbs being used in the treatment of diseases and for revitalizing various body systems in almost all ancient civilizations. Plants have traditionally served as man's most important weapon against pathogens. Herbal medicines are widely used by all sections of community either as folk remedies or as medicaments in the indigenous as well as modern system of medicine. The tea tribes is one of the ethnic group of Assam, have great knowledge about their traditional medicines for their different types of diseases. The present study has been conducted to record the medicinal plants that are used by the tea tribe people of Upper Assam for the treatment of viral Hepatitis. Two different Tea estate of Tinsukia district of Assam has been selected for the study. The methods employed in the study are survey, observation, interaction with the tea tribe people and discussion with the Kabiraj/Ujha and Medical Officers. More than 30 different types of medicinal plants has been collected from the study area and reported in this paper along with their botanical name, family, useful parts and mode of application. The present study aims to draw the attention of researches towards the need of future critical study.

Keywords: Traditional phytomedicine, Tea tribes, Hepatitis.

Introduction

India is sitting a gold mine of well recorded and traditionally well practiced knowledge of herbal medicine. Recently considerable attention has been paid to utilized eco friendly and bio friendly plant based products for prevention and cure of different human diseases. Hepatitis is an inflammatory disease caused by infection of viruses in liver cells. Many thousands of people die each year in the world due to Hepatitis, out of which HB+V (Hepatitis B Virus) is more serious and lead to liver sirosis and cancer. As per world Health Organization report 2003, 5000 people die each year in USA, due to HB+V. Hepatitis can be caused by alcohol, contaminated food, water and milk, poor sanitation and fecal oral route transmission. Hepatitis can be caused by half dozen viruses like Hepatitis A, Hepatitis B, Hepatitis C, Hepatitis D, Hepatitis E and Hepatitis G viruses. But Hepatitis A and B are most common in our country. Hepatitis A is commonly known as epidemic Jaundice.It is an infectious disease caused by Hepatitis A virus (HA+V). The disease is heralded by non specific symptoms such as fever, chills, headache, generalized weakness, dark aches and pain followed by dark yellow urine and jaundice. Hepatitis B is an acute systemic infection with major pathology in the caused by Hepatitis virus (HB+V).

Medicinal plants play a vital role in all system of medicine. Near about 50 per cent effective medicine of the world are extracted from plant sources which are traditionally using in various system in medicine. The use of herbal medicine in the treatment of liver disorder has been in tradition from long back. The first literature to have reported on it was *Charaka Samhita* in 600 BC. The important of herbal medicine in the treatment liver disorder can be traced back to 2100 BC.

In Indian Ayurvedic system of medicine more than 100 plant species are commonly used as therapeutic drugs for control and prevention of liver diseases. A vast knowledge of ethno-medicinal plants exists as oral among the ethnic communities and tribes in North East India are silently losing with time having its significant therapeutic value. The tea tribes, one of the ethnic group of Assam who have their origin from chotanagpur, Madhya Pradesh and Orissa have great knowledge about their traditional medicine. In Assam the tea tribes are generally from the tribes of Munda, Orang, Ghasi, Tantis, Lohar, Sonar, Goala, Nagbansi, Mahali, Sawra etc. Literacy among tea tribes is very poor and they are economically backward.The knowledge of health and hygiene also very poor among the tea tribe people and they are generally affected by some frequently occurring common diseases like diarrheoa, dysentery, malaria, typhoid, worm disease, skin diseases, tuberculosis etc. Like all other communicable diseases, Hepatitis A also commonly seen among the tea tribe people due to consumption of contaminated food, water, alcoholic beverage etc. It have been observed that the tea tribes used different types of medicinal plants for Hepatitis treatment as food supplements and therapeutic drugs.

Through the study it is trying to highlight the recent status of traditional treatment for Hepatitis diseases among the tea tribe community of Assam. The present work have also aims to explore the herbal medicine available in the study area for the treatment of viral Hepatitis and to provide information that will be useful to phytochemists and pharmacologists for further exploration.

Area of the Study

The area for the present study covers the Tinsukia district of Assam. Tinsukia district is known for its tea gardens and natural resources. Total number of tea gardens in Tinsukia district is 120. The district is extended from 27°23' north to 27°48' north latitude and 93°22' east to 95°38' east longitude. Total geographicl area for the district is 3790 Sqkm. Talap Tea estate and Khobang Tea estate situated approximately 13 kilomiters away from Doomdooma town of Assam. Talap Tea estate is considered as "The Golden Belt of Assam". This estate derived its name "Talap" which in Hindi means lakes, ponds and water bodies. Talap Tea estate was formed after amalgamation with Dangri tea estate and one part amalgamates with Khobong tea estate known as Kherbari division.

Objectives

1. To find out the main reasons of the disease.
2. To find out what type of treatment they prefer, traditional or modern?
3. To assume whether they get better result from their traditional medicine or modern medicine?
4. To find the ethnic herbal plants used for the treatment of Viral Hepatitis.

Methodology

Two different tea estates of Tinsukia district had been selected for the study of traditional phyto medicine used for Hepatitis treatment among the tea tube community of upper Assam. The two tea estates are:

1. Talap Tea estate Appejay group
2. Khobang Tea estate Appejay group

The methods employed in the study were field visit, survey, observation, interaction with the tea tribe people and discussion with the Kabiraj/ujha and medical officers.

Two Separate type of questionaries were designed one for the tea tribe people and one for the Kabiraj/Ujha to have depth information about their traditional treatment for Viral Hepatitis.

The entire work had done on the basis of primary data, collected from Kabiraj/Ujha, tea tribe people and from medical officers. More than 30, different type of plant species had been collected and identified. Photograph of a few plants had also taken during the time of field visit and compared with ethno botanical literature (A Handbook of Medicinal Plant, Prajapati Narayan Das *et al.*).

Results

During the study a survey had been carried out among the 100 families of tea tribes people. Observation discussion and interview had been carried with 16 traditional medicine men/women and a separate discussion had also been carried with the medial officers of Talap and Khobong tea estates. Followings are the findings of the study.

Reason of the Disease

From survey it had been find out that the main reasons of viral Hepatitis among the tea tribe people are 1. Consuming contaminated food 2. Consuming contaminated water 3. Living unhygienic crowded condition 4. Consuming alcoholic beverages 5. Improper sanitation

Consuming Contaminated Food

Consuming contaminated food is one of the major factor for viral Hepatitis. The tea tribe people have not any knowledge about the food hygiene and the food in their daily diet are very inferior quality. They have taken the uncovered food prepared on previous day. They wash their food with contaminated water and do not cooked the food properly So consuming raw and inadequately cooked food is also caused the viral Hepatitis

Consuming Contaminated Water

Most of the tea tribe people do not purify the water before they drink. Filtrations have done by very few families. They generally drink contaminated water directly from the tube well or from the tap provided by the tea garden authority. So consuming contaminated water is another cause of viral Hepatitis among the tea tribe people.

Improper Sanitation

Although the tea garden authority have given a good facility for sanitation by providing the sanitary and borehole latrine for tea garden people but the tea tribe people do not like to defected inside the latrine. They like to defecate outside particularly inside the tea bushes, in open field or near the river. They like to use the latrine to keep their poultry an fuel wood. Fecal contamination is another main cause of viral Hepatitis, among the tea tribe people.

Living Unhygienic Crowded Condition

The tea tribe people do not maintain the family planning measure. They like to live in a large family which is generally unhygienic. They live in small pucca or kachcha houses provided by the tea garden authority. Hepatitis is a infectious disease, it can spread by direct contact with person to person. Since the tea tribe people live a large family in a small house, so it can spread easily from person to person.

Consuming Alcoholic Beverages

Drinking liquor is very common among the tea tribe people. They generally drink their home made beer *i.e.* Haria or Laupani. Sometimes they also drink the country liquor. Male worker drinks more than the female worker. Drinking alcoholic beverages is another main cause of Hepatitis disease among the tea tribe people.

Traditional Treatment and Traditional Medicine Men (Kabiraj/Ujha)

The tea tribe people have great faith on their traditional treatment. They always prefer their Traditional medicine in case of all type of diseases including viral Hepatitis. They only go for modern treatment when their traditional medicine fail to cure the diseases.In case of Hepatitis disease the tea tribe people always give the first preference to their traditional medicine. Around 90 per cent people had given their statement that Hepatitis can be cured by traditional medicine.

The tea tribes have two different type of traditional treatment. (i) Herbal treatment (ii) Supernatural treatment. 70 per cent people prefer their Herbal treatment and 30 per cent people prefer the supernatural treatment for viral Hepatitis. The person who treats the disease with herbal medicine is called Kabiraj and the person who treats the disease by their supernatural power is called Ujha. Interview had taken from 16 Kabiraj/Ujha from two different tea estates. Out of 16 traditional medicine men 14 are Kabiraj and only 2 are Ujha. 8 Kabiraj and 2 Ujha are from Talap T.E. and 6 Kabiraj from Khobong T.E. Most of the Kabiraj/Ujha had learned their knowledge either from their ancestor like Father/Grandfather or from their Guru. Most of the kabiraj are old and they have more then10-30 years experience about their traditional treatment. AII most all the kabiraj conserve the herbal plants in their kitchen garden. The lists of kabiraj/ojha who treat the viral hepatitis have given in Table 14.1.

Educational Status of Traditional Medicine Men (Kabiraj/Ujha)

Educational status of traditional medicine men is very poor among the tea tribe community. Out of 18 traditional medicine men only one kabiraj have intermediate qualification, rest have below matriculate qualification. The Figure 14.1 has represents the educational status of traditional medicine men of tea tribe people.

Table 14.1: List of the Traditional Medicine Men who Treat the Hepatitis Disease

Sl.No.	Name of the Kabiraj/Ujha	Name of the Garden	Educational Qualification	Occupation
1.	Raikha Munda (Kabiraj)	Talap T.E.	Nil	Temporary Worker
2.	Jagdevi Kumri (Kabiraj)	Talap T.E.	Class VI	Permanent Worker
3.	Sabitri Devi (Ujha)	Talap T.E.	Class III	Non Worker
4.	Sivasagar Prasad (Kabiraj)	Talap T.E.	Class IX	Permanent Worker
5.	Kapil Bhokta (Kabiraj)	Talap T.E.	Class VI	Temporary Worker
6.	Arjua Karmakar (Kabiraj)	Talap T.E.	Nil	Permanent Worker
7.	Gobai Munda (Kabiraj)	Talap T.E.	Class III	Temporary Worker
8.	Logiri Baraik (Ujha)	Talap T.E.	Nil	Non Worker
9.	Lucas Kalondi (Kabiraj)	Talap T.E.	Nil	Non Worker
10.	Dasai Munda (Kabiraj)	Talap T.E.	Nil	Temporary Worker
11.	Lakhan Das (Kabiraj)	Khobong T.E.	Class VI	Non Worker
12.	Iswar Gwalla (Kabiraj)	Khobong T.E.	Class VII	Temporary Worker
13.	Nirmal Bharjo (Kabiraj)	Khobong T.E.	Class IX	Permanent Worker
14.	Mangal Rawita (Kabiraj)	Khobong T.E.	Class V	Permanent Worker
15.	Muri Banik Das (Kabiraj)	Khobong T.E.	Class I	Permanent Worker
16.	Navin Das (Kabiraj)	Khobong T.E.	Class II	Temporary Worker
17.	Vijay Rawtia (Kabiraj)	Khobong T.E.	Class XII	Permanent Worker
18.	Rajaram Munda (Kabiraj)	Khobong T.E.	Nil	Temporary Worker

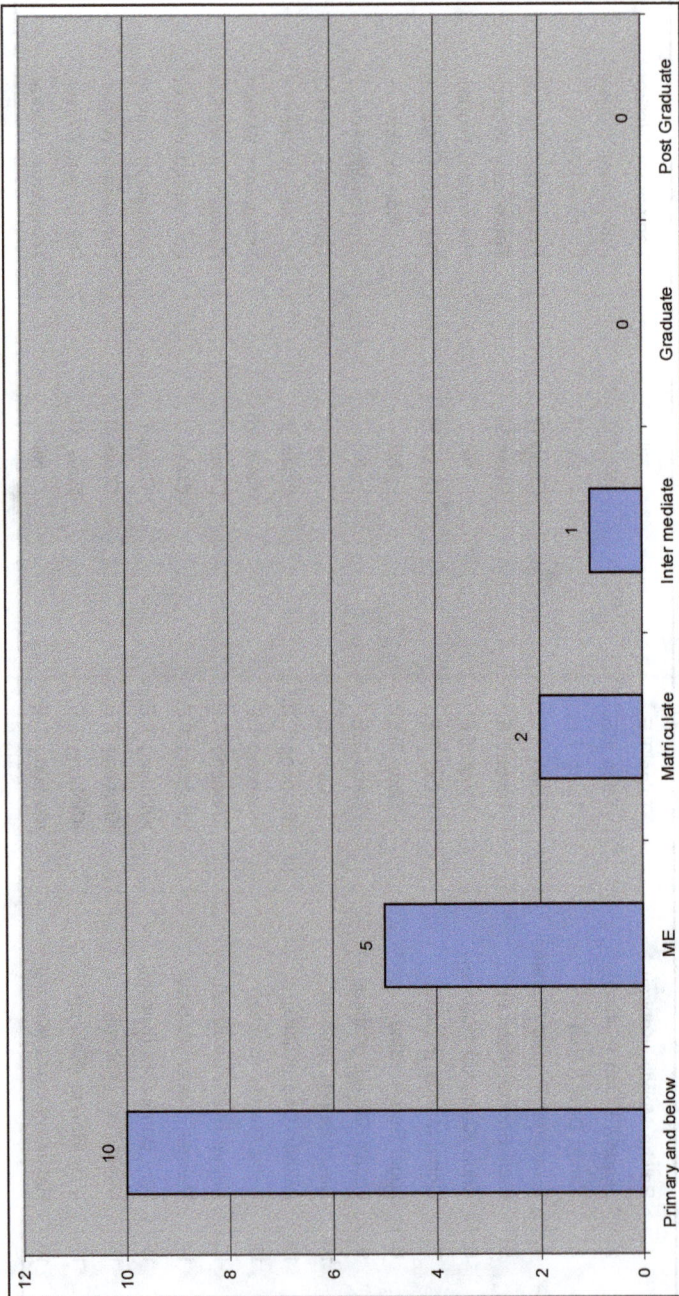

Figure 14.1: Educational Status of Traditional Medicine Men (Kabiraj/Ujha)

Table 14.2: Ethno-medicinal Plants Used by the Tea Tribe People to Cure the Hepatitis Disease

Sl.No.	Scientific Name	Vernicular Name	Family	Useful Plants(s)	Method of Preparation	Mode of Application
1.	*Ricinus communis*	Aren pat	Euphorbiaceae	Leaves	Juice extracted from leaves	Taken internally, 1 tea spoonful twice daily till 7 days
2.	*Vinca rosea* *Dalbergia sissoo*	Sadabahar Sissoo	Apocynaceae Fabaceae	Leaves Leaves	Mix the leaves extract of *Vinca rosea* and *Dalbergia sissoo* with missiri and make it paste with water	2 glass daily till 7 days
3.	*Tinospora cordifolia* *Hydrocotyl rotundifolia* *Piper nigrum*	Titakhori Sonamoni Gulmarich	Tinosporaceae Umbelliferae Piperaceae	Root Whole plant Seeds	Boil 10 gm of root of *Tinospora cordifolia* in ½ lit. of water with juice of *Hydrocotyl rotundifolia* and make paste with missiri and 5 gm of seed of piper nigram	1 glass 3 times per day till cure
4.	*Costus specious* *Musa paradisica*	Bon kohiyar Chini kol	Zigiberceae Musaceae	Root Root	Mixed the root of *Musa paradisica* and *Costus specious* and make powder with missiri and make paste with water	1 glass per day till 7 days
5.	*Terminalia arjuna*	Arjun Milong guti	Combertaceae	Leaves Seeds	Mix the leaves crushed with the powder of Milonguti and Talmissiri and make it liquid with 9 bottles of water	Given internally 1 bottle per day till 9 days.

Contd...

Table 14.2–Contd...

Sl.No.	Scientific Name	Vernicular Name	Family	Useful Plants(s)	Method of Preparation	Mode of Application
6.	Mangifera indica	Aam	Anacardiaae	Bark	Mix the bark of Mangifera indica with caco₃ and keep with overnight with 1 ltr water in a pital container	Given externally from head to toes by Ujha with some special Mantra doing 3 times in a day
7.	Mentha arvensos	Pudona	Lamiaceae	Leaf	Crushed and juice prepared. Leaves are put in water for 6-8 hrs, decanted extract prepared.	Taken internally 2-3 tea-spoonfull in a day for 7 days after breakfast
8.	Phyllanthus niruri	Jangli Auora	Euphorbiceae	Whole plant	Crushed into paste and juice prepared	Taken internally 2 tea-spoonfull 3 times a day.
9.	Litsca glutinosa	Agyachi	Lauraceae	Root	Juice extract from root	Taken internally 2-3 tea-spoonful 2-3 times per day.
10.	Leucars aspara Spilanthes paniculata	Durun sak	Lamiacea Astreaceae	Leaves Inflorencence	10-15 no. of leaves of Leucas aspara mix with 3 matured inflorences of Spilanthes peniculata crushed into pills.	Taken internally 1 pill thrice a day after food

Contd...

Table 14.2–Contd...

Sl.No.	Scientific Name	Vernicular Name	Family	Useful Plants(s)	Method of Preparation	Mode of Application
11.	Terminalia arjuna Azadirachta indica Sida cordifolia Spilanthes paniculata Sida spinosa Desmodium laburni-folium	Arjuna Gas Maha neem Sirbioral	Combretaceae Meliaceae Malvaceae Astreaceae Malvaceae Fabaceae	Bark Bark Root Root Root Root	100 gm of bark of *Terminalia arjuna* and *Azadirachta indica* and 50gm of root of each of the rest taken crushed and boiled with water about ½ liter and missiri powder.	1 glass in empty stomach
12.	Carica papaya Cyperus brevifolius Musa bulbisiana	Papita Ghass Gutikol	Caricaceae Cyperaceae Musaceae	Root Rhizome Root	50gm of root of each *Musa balbisiana* and *Carica papaya* and equal amount of rhizome of *Cyperus breivifolius* crushed and dipped overnight in above ½ liter of water extra separated and stored in	1 cap daily in empty stomach
13.	Averrhoea carambola	Kamarangah	Amranthaceae	Fruit	Prepared the juice from ripen fruit	1 glass per day till the diseases cure
14.	Lowsonia innerimis	Mehendi	Lythraceae	Bark	About 30 gm of bark is boiled with 1 ltr water and cooled separate the extract and stored it	1 cup daily in empty stomach
15.	Alstonia scholaris	Soitan	Apocynaceae	Bark	Bark cut into small pieces and make necklace and given to the patient	Externally as necklace

Contd...

Table 14.2—Contd...

Sl.No.	Scientific Name	Vernicular Name	Family	Useful Plants(s)	Method of Preparation	Mode of Application
16.	*Ananas comosus*	Matikathal	Bromeliaceae	Leaves	Extract the leaves juice	Taken internally 2-3 tea spoonful 3 times daily
17.	*Oldanlandia corymbosa*	Jangli Gul marich	Rubiaceae	Whole plant	Extract the juice from the whole plant	Taken internally 2-3 teaspoonful per day
18.	*Premua latifolia*	Agnimantha	Verbenaceae	Bark	Bark doctrine is prepared	Taken internally
19.	*Alstonia scholaris*	Sotiana	Apocynaceae	Bark	Fresh bark extract in water and kept overnight	Taken internally 1 cup per day in empty stomach
20.	*Bryophyllum pinnatum*	Pategaza	Crassulaceae	Leaves	Leaves juice extract	Taken internally 2-3 spoonful 3 times per day
21.	*Paederia foetida*	Padrung pat	Asclepiadaceae	Root	Fresh root grind and mix with water	Taken orally
22.	*Azadirachta indica*	Neem	Meliaceae	Bark	25 gm powder of bark is added to water and mix with missiri	Taken daily for seven days to cure the disease
23.	*Allium cepa*	Pyaj	Liliaceae	Bulb	50 gm bulb of onion cooked with vinegar	1 tea spoon full 3 times per day
24.	*Justicia adhotoda Plumbago zeylanica*	Adusa	Acanthaceae	Leaves	100 gm powder of *justicia anhotoda* mixed with leaf powder of *Plumbago zeylanica* mix with milk	2 spoon daily 3 times till 7 days

Contd...

Table 14.2–Contd...

Sl.No.	Scientific Name	Vernicular Name	Family	Useful Plants(s)	Method of Preparation	Mode of Application
25.	*Cassia fistula*	Girimala	Caesalpiniaceae	Fruit pulp	Fruit pulp drink and make past with water	25 gm fruit pulp twice a day upto 2 weeks
26.	*Emblica officinalis*	Auora	Euphorbiaceae	Fruit and seeds	Whole fruit grind and make it powder	50 gm powdered drug is given thrice day with milk
27.	*Phyllanthus niruri*	Jangli aura	Euphorbiaccae	Fruit	100 gm of fruit grind and make paste with milk	1 times a day
28.	*Lagenaria vulgaris*	Loki	Cucurbitaceae	leaves	Leaves juice extract and mix with water	1 cup per day
29.	*Tinospora cordifolia*	Saguni lota	Tinosporaceae	Root	Grind The root and make it powder and mix with water and missiri	50 ml taken 2-3 times per day.
30.	*Aloe barbadensis*	Katapat	Liliaceac	leaves	Extract the leave juice and mix with black salt and ginger and make paste	2-3 spoon 3 times per day till 20 day
31.	*Cuscuta reflexa*	Akashlata	Convolvulaceae	Stem	Extract the juice of stem and mix with missiri powder	2-3 teaspoonful 3 times per day
32.	*Mimosa pudica*	Lajoni kata	Mimosaceae	Root	Grind the root and mix with water	2 spoonful 2 times per day
33.	*Curcoma longa*	Haladhi	Zingiberaceae	Rhizome	Extract the juice of rhizome and mix with milk	1 glass per day

Contd...

Table 14.2—Contd...

Sl.No.	Scientific Name	Vernicular Name	Family	Useful Plants(s)	Method of Preparation	Mode of Application
34.	Saccharum officinarum	kohiyar	Graminaceae	stem	Keep the sugar can piece open whole night and allow the dew drops fall on it	Chew the kohiyar in the morning in empty stomach till four days
35.	Aegel marmelos	Bel	Rutaceae	leaves	Extrack–10-30 gm of the leaf juice and add ½ gm of black paper powder	Take twice a daily
36.	Syzygium aromaticum	Long	Myrtaceae	Bud	Soak 10-15 cloves in water at night and crushed next morning and sieve	Drink 1 cup on empty stomach
37.	Momordica charantia	Tita kerela	Cucurbitaceae	leaves	Extract the 10-15 ml of juice of leaves and mix with missri	2-3 tea spoonful twice a daily

Ethno-medicinal Plants Used by the Tea Tribe People for Hepatitis Treatment

The traditional medicine men used more than 30 medicinal plants for the treatment of viral Hepatitis in tea tribes community. The medicinal plants used for Hepatitis treatment enlisted in Table 14.2 along with their scientific name, family, useful parts and mode of application among tea tribe people.

Comparative Study Between the Preference of Traditional and Modern Treatment

A comparative study between the preference of traditional and modern treatment for Hepatitis disease had also done among the tea tribe people.

In Talap T.E. 90 per cent people prefer the traditional treatment and only 10 per cent people prefer the modern treatment for viral Hepatitis.

In Khobong T.E. 87 per cent prefer the traditional treatment and 13 per cent people prefer the modern treatment for viral Hepatitis. The comparative study have shown in Table 14.3.

Table 14.3

Sl. No.	Name of the Garden	Preference of Traditional Treatment	Preference of Modern Treatment
1	Talap T.E.	90 per cent	10 per cent
2	Khobong T.E.	87 per cent	13 per cent

From the above observation it had been find out that the tea tribe people do not prefer the modern treatment in case of viral hepatitis, they generally prefer their traditional health care even though they have free medical facilities.

Discussion

In the north east India, the tea garden areas have lots of medicinal plants as wild condition. The tea tribe people generally have poor knowledge about the health and hygeine and they are generally affected by some frequency occuring diseases. Hepatitis A, i.e epidemic Jaundice also has commonly seen among the tea tribe people. The tea tribe people always prefer their traditional medicine in all type of illness. In case of hepatitis they always prefer either herbal or supernatural treatment but only in serious cases they go for modern treatment (when their traditional medicine fail to cure the disease).The traditional practitioner never demand too much money from their patient. They treat their patients by observing symptoms of the patients, reading pulses doing puja's etc. It is also believe in the tea tribe community that if the *kabiraj* or *ojha* demand too much from their patient then their herbal/supernatural knowledge would be taken away by God who on sure his/her with special gift. Though the knowledge of educational practitioner is very poor but they treat the patient with their old experiences and knowledge of basic norms of ethno-medicine viz toxic free site selection, maturity of the plant and it parts used, and knowledge of experiences of raw drug behaviour. Most of the kabiraj conserve the medicinal plants in their kitchen garden (bari) without having proper knowledge of conservation. Lots of medicine plants available in the study area most of them are found in wild condition. Out of these 38 species use for hepatitis treatment have been collected and indentified under with the help of herbarium of Namrup College.

Different parts of the parts are used as herbal medicine for hepatitis treatment and the mode of application of administration are paste, juice, infusion powder etc. and in some cases they used salt, sugar, missiri, honey etc. for changing base or test of the medicine.

The reported species have protects the liver in many ways with boostering immune system, cleaning the blood bacteria and wash product maintaining has more balance producing quick energy on demand, helping in production of note etc

Conclusion

The traditional uses of plants by different tribes may be regarded as the basic material for scientific documentation studies. Therefore in the last few years' traditional knowledge of plants come into focus for research work and for development of other value added product.

In the study area though the people are educationally backward but the knowledge of herbal treatment is more among the tea tribe community. Step should be taken to educate these people, so that they can develop their knowledge scientifically for their herbal treatment. Instead of this, step should also taken in conservation of medicinal plant in community level, campaign, focusing through mass media and small scale cultivation at family level for quality best production.

After investigation it have been conclude that medicinal plants are giving to play a very important role in helping those frustrated by the modern allopathic medicine and suffering from its grave side effect, that is why the tea tribe people always looking back to nature for safety and security. Herbal medicines are cheaper, easily available and mode of preparation is also simple. There is no risk of any major side effects and above all it suits the social and culture needs of people. These medicines also observe evaluation or modern scientific lines such as play to chemical analysis, biological screening, pharmacological investigation and chemical trials for exploring new lead compounds for treating the elements faced by the mankind nowadays.

References

Bhadra, R.K., 1997. *Social Dimension of Health of Tea Plantation in India.* N.L. Publishers, Dibrugarh, Assam.

Biswas, J.K., Das, A., Shaw, G.C., Mukherjee, N. and Chakrabotary, M., 1990. Studies on outbreak of viral hepatitis at Calcutta with special reference to serological investigation. *J. Ind. Med. Ass.,* 88(9): 257–259.

Dubey, N.K., Kumar, R. and Tripathi, P., 2004. Global promotion of herbal medicine, India's opportunity, *Cur. Sci.,* 861(1): 37–41.

Flora, K.D., Rosen, H.R. and Brenner, K.G., 1996. The use of neuropathic remedies for chronic liver diseases. *Am. J. Gastroenterol.,* 91: 2654–2658.

Gupta, 1986. Tribal concept of health diseases and remedy. In: *Tribal Health Socio Cultural Dimension,* (Ed.) B. Choudhary. Inter-India, New Delhi.

Hasan, K.A., 1979. *Medical Sociology of Rural India.* Sachin Publication, New Delhi.

Mital, K., 1979. Primitive medicine verses modern medicine among the santhals. *Journal of Social Research,* 22(1).

Park, K., 2002. *A Textbook of Preventive and Social Culture of Medicine.* M/s Banarsidas Bhanot Publishers, Premnagar, Jabalpur, India.

Prajapati, Naroyan Das, Purahohit, S.S., Sharma, Kumar Arun and Kumar, Tarun, 2003. *A Handbook of Medicinal Plant.* Agrobios, Rajasthan.

WHO, 1988. World Health Organisation, May.

Chapter 15

Use of Medicinal Plant for the Treatment of Diabetes and their Importance in our Rural Economy

Indira Baruah

ABSTRACT

Plants have played an important role in medicine. It is estimated that 80 percent of the world's population depends directly on plant based medicine for their health care WHO (2003). India is characterized by a wide range of climatic, geographical and geological conditions within which infinite varieties of many rare, precious herbs and trees flourish. Now, it is globally recognized that traditional alternative complementary system of medicine provide health to the population of both developing and less developed countries. Even in the developed world realize that the modern system of medicine alone is not able to address all the health problems, particularly related to the life style disease like diabetes, hypertension, AIDs, cancer etc. It is in this context we can say that traditional health care system can be used to find solutions to the present day problems. An attempt is made to know about the use of medicinal plants for the treatment of diabetes, which is recognized as one of the most dreadful diseases. This paper highlights the medicinal plants used for the treatment of diabetes and their importance in rural economy by assessing the market potential of the recorded plants. In the investigation, 20 ethno-medicinal plants have been recorded for the treatment.

Introduction

Historically, plants have played an important role in medicines. Through observation and experimentation, human beings have learnt that plants promote health and well being. The use of medicinal plants for health reason started thousands of years ago and is still part of medicinal practice in china, Middle East, Africa, Egypt, South America, India and other developing countries. Over the centuries the use of medicinal herb has become an important part of daily life despite the progress in modern medical and pharmaceutical research. The *Rigveda* (3700 BC) mentions the use of medicinal plants.

The herbs and medicinal plants, which grow naturally in the garden and open areas, have the power of healing diseases over traditional system of medicine *i.e.* Ayurveda, Unani, Homeopathy etc. use herbs for treatment. Ayurveda contains details about these natural medicinal plants. It also contains the description about the art of recognizing these plants and their healing qualities. During olden days people used to use the herbs and shrubs, which were easily available in the nature and were full of medicinal qualities. Today the herbs and shrubs are once again being used in all the houses, whether rich or poor and their utility is increasing day by day. It is estimated that 80 per cent of the world populations depends directly on plant based medicine for their health care (WHO 2003). It is now globally recognized that traditional alternative complementary system of medicine provide health care to almost 80 per cent of the population of developing and less developing countries. Even in the developed countries, realized that modem system of medicine alone is not able to address all the health problems, particularly related to the life style diseases like Diabetes, Hypertension, AIDS, Cancer etc. In India, medicinal plants offer low cost and safe health care solution. This is an attempt to know about the use of medicinal plant for the treatment of diabetes and their importance in our rural economy.

Aim and Objectives

1. To study the use of medicinal plant for the treatment of diabetes.
2. To study the importance of these medicinal plant in our rural economy.

Methodology

This paper is based on both primary and secondary data. For primary data Interview was taken with the help of interview schedules, containing structural questions and also through telephonic conversation. Secondary datas are collected from books, magazines, newspapers etc. For the study 50 (fifty) respondents were selected who have been suffering from diabetes. The age group of the respondents was from 20 to 80 years (Table 15.1). Among them, most of the respondents (34 per cent) were from the age group of 40-50 years (Table 15.1).

Table 15.1

Sl.No	Age	No of Respondent	Percentage (per cent)
1.	20-30	02	4
2.	30-40	10	20
3.	40-50	17	34
4.	50-60	15	30
5.	60-70	04	8
6.	70-80	02	4
	Total	**50**	**100**

Table 2

Sl.No	Types of Respondents	No of Respondents	Percentage (%)
1	Insulin dependent	06	12
2	Non-Insulin dependent		
	a. Depends only on tablets	20	40
	b. Both tablets and medical plant	22	44
	c. Depends only on medical plant	2	4
	Total	**50**	**100**

The respondents were classified into two groups (Table 15.2):

1. Insulin dependent
2. Non-Insulin dependent

Result

Altogether twenty three medicinal plants were recorded which have been used by diabetic patients of the study area (Table 15.3). These plants are easily available in the locality and hold definite promises in the management of diabetes. The study indicated significant activities of the extracts of different medicinal plants as antidiabetic and support traditional usage to prevent diabetic complications. Inspite of the presence of known antidiabetic medicine in the market, remedies from medicinal plants are used with success to treat this crippling disease.

Discussion

Mostly the knowledge and information were passed from generation to generation by word of mouth and through folklores. The most important aspect is the method of preparation and mode of administration that must be accurate and clean otherwise there might be some side effects. The respondents in this study area use different medicinal plants which are easily found in their garden and nearby forest. The medicinal plants recorded for treating diabetes has potential market value. A good number of such plants are found to be sold in the local market. There could be greater possibility of wider market potentialoity of some plants which can be easily cultivated in the area such as Aloe vera, Nayantora, Stevia, Amla. Nevertheless, these plants are also used against different ailments other than diabetes.

The people of this area have not been able to explore the market potential of these medicinal plants. Lack of awareness, lack of knowledge about cultivation practices and processing as well as market facility are some of the reasons behind this scenario. Through proper training and demonstration, the farmers of this area should be encouraged to cultivate valuable medicinal plants.

Conclusion

The rural communities especially in remote areas rely greatly on age-old traditional knowledge for health care. The medicinal

Table 15.3: Medicinal Plants Used by the Respondents Against Diabetes

Sl.No.	Name of the Plant	Botanical Name	Method of Preparation and Administration
1.	Amla	*Phyllanthus emblica*	(a) Take 1 tea spoon Amla Powder with a cup of water twice a daily.
			(b) Juice of fresh Amla can be taken regularly.
2.	Tejpat	*Chinamomcem tamla*	Soak three cinamomam leaves with water over night and take next morning after meal
3.	Periwinkle (Nayantora)	*Catheranthus roseus*	Take two periwinkle fresh leaves and use only three days in a week after meal
4.	Garlic	*Allium sativam*	Fresh garlic can be eaten daily
5.	Bitter gourd	*Momordica charantia*	(a) Consume juice of fresh fruits every day.
			(b) Grind shade dried fruits to the powder and take three-five gm of powder with water every day.
			(c) Take a glass of juice of fresh leaves every day morning on empty stomach
6.	Black berry	*Syzygixium cuminii*	(a) Powdered seeds can be taken twice daily.
			(b) Grind dried fruits with the seeds to make powder. One tea spoon of this powder can be taken.
			(c) Black berry juice can be used.
7.	Neem	*Azadirachta indica*	(a) Dried leaves can be used.
			(b) One tea spoon of Neem powder can be taken with a cup of water after light meal.
8.	Arjun	*Terminalia arjuna*	(a) Boil the bark with water, and take in the morning in empty stomach
			(b) Make powder from bark of Arjuna after drying in sunshine and one spoon powder can be boiled with a cup of milk and can be used.

Contd...

Table 15.3—Contd...

Sl.No.	Name of the Plant	Botanical Name	Method of Preparation and Administration
9.	Honey plant (Mau Tulsi)	*Stevin rebaudiayana*	(*a*) Both fresh and dried leaves can be used. Five fresh leaves of stevia can be eaten. (*b*) Powder can be made after drying in sunshine 1 tea spoon of stevia powder can be mixed with a cup of water and take twice daily.
10.	Indian aloe	*Aloe barbedensis*	A small piece of aloe vera can be eaten.
11.	Ban Amlakhi	*Phyllanthus fraternus*	Juice can be made from leaves and stems. Half cup of juice can be eaten regularly.
12.	Atlas	*Annona squamosa*	Make small pieces of bark of the plant after drying in the sunshine. Two or three pieces are boiled with three cups of water to make a half cup. Half cup juice can be taken.
13.	Wheat	*Triticum aestivum*	Remove the wheat plants from the roots and cut the roots with the help of scissors and grind the wheat plant on the grinding stone and keep adding water to grind it to a fine paste. Consume fresh juice of wheat at least 50ml.
14.	Brahmi	*Bacopa monnieri*	The juice of fresh leaves (15-30 ml) may be taken.
15.	Fenu greek (methi)	*Trigonela foenum graecum*	(*a*) Grind the seeds to fine powder. Take 3-5 gm of this powder with water twice daily. (*b*) Soak one tea spoon seeds in water over night. In the morning drink the water and eat the seeds empty stomach.
16.	Bael	*Aegle marmelos*	Juice of fresh leaves can be taken in the morning on empty stomach.
17.	Cinnamom, dalchini	*Cinnamonum zeylanicam*	Soak two or three pieces of cinnamon in a cup of water and take it in the morning in a empty stomach
18.	Curry leaf	*Murraya koenigi*	Curry leaves can be used to reduce blood sugar.

Contd...

Table 15.3—Contd...

Sl.No.	Name of the Plant	Botanical Name	Method of Preparation and Administration
19.	Small pennywort	*Hydrocotyl sibthnopioides*	Three spoons of Manimuni juice can be eaten in the morning on empty stomach or take it twice a daily
20.	Leaves of rahar dal	*Cujanas cajan*	½ cup of juce can be used
21.	Kunduli		(*a*) Boiled kunduli leaves can be eaten (*b*) Kunduli can be eaten as a salad and vegetables
22.	Pomegranate	*Punica granatum*	6-7 leaves of pomegranate can be eaten regularly
23.	Fig (Dimoru)	*Ficus glomenata*	Juice made from fig with a glass of water can be taken in the morning in empty stomach

plants which are used for the treatment of diabetes are safe and inexpensive. Now, time has come to good use of traditional knowledge of medicinal plants through modern approaches of drug development. By using medicinal plants and practicing regular physical exercise the level of blood sugar can be maintained. Cultivation of antidiabetic plants may not only improve local economy but also helps to conserve and manage the bio-diversity of the region.

References

Kurukshetra, 2007, 56(2): 46–48, December .

Kurukshetra, 2007, 55(9): 26–27, July.

Kurukshetra, 2007, 55(4): 46–48, February.

Kurukshetra, 2007, 57(9): 44–47, July.

Kurukshetra, 2008, 56(5): 18–22, March.

Pandey, B.P. *Economic Botany.* S.Chand Company Publication Ltd., New Delhi.

Yoga Sandesh, 2005, (3): 46–49, November.

Chapter 16

Use of Non Timber Forest Products for the Upliftment of Rural People with Special Reference to Arunachal Pradesh

Pronab Mudoi, Parkkal Rethy and Binay Singh

ABSTRACT

Non Timber Forest Products (NTFPs) are the most important resource for upliftment of economy of rural people. In recent years NTFPs have attracted considerable global interest as it able to provide important community needs for the improvement of rural population. Indian rural and tribal economy depends basically upon forest produces. NTFPs play a key role in the life and economy of rural and tribal communities living in and around forests. Arunachal Pradesh located in extreme north eastern corner of the country is endowed with a rich source of non timber forest products (NTFPs) distributed throughout the length and breadth of the state. In this state the usages of NTFPs are high as because the maximum population is dominated by tribes which are directly or indirectly depend upon forest products for their livelihood.

Introduction

All forest products other than industrial timber, which are being used for the welfare of the society are considered as Non Timber Forest Products. The non timber forest products include food, medicines, bamboos and canes, fiber, tans and dyes, oilseeds mushroom etc. The maximum tribal people in developing countries depend on forest products to fulfill their nutrition, health, house construction or other needs. Most of the non urban people below the poverty line almost entirely depend on forests for their existence. It has been estimated that NTFPs generate about 1.2 million employees per year, which is about 55 per cent of the total employment generated in the forestry sector (Dwivedi, 1993). Theagarajan, 1994 also reported that NTFPs provide about 50 percent of income for 20-30 percent of the rural and tribal people in India.

Arunachal Pradesh is an important Vavilovian Centre of diversity and origin of important cultivars species and some domesticated animals (Baishya *et al.,* 2002). It is a vast repository of plant resources with both ecological and economic importance. The resource value of this region spans from non timber category, medicinal and aromatic to food and industrial gene pools and other economically important sectors. It comprises well over 6000 species of flowering plants of which nearly 30-40 per cent is endemic to this region (Roy and Behera, 2005). The economic growth of people of the state depends mainly on forest resources.

Rural development involves a process of developing the rural economy by NTFPs to raise the standard of living of those rural people who are poor and require upliftment.

Use of NTFPs for Human Welfare

Medicinal plants have great economic potential for rural people as it will open up unexplored avenues in cultivation, propagation, processing, packaging and marketing of medicinal plants. Medicinal plants are easy to grow. The tribal people in rural areas utilize different types of NTFPs for fuel, fodder, food, housing building, medicines etc. About 1,800 medicinal plants have been documented in Ayurveda literature alone.

The World Health Organization (WHO) reports that 80 per cent of the population of developing countries depends on traditional medicines, mainly plant drugs. In India, one seventh of the total plant species yield NTFPs such as fruits, seeds, resin, fiber, bamboo, gums, oils, tans, dyes and medicinal plants (Table 16.1). Arunachal is one of the mega biodiversity hot spots region in India. The state is rich in drug yielding trees, shrubs, herbs and climbers. More than 500 species of medicinal plants and herbs are growing in different agro-climatic zones of the state (Haridasan *et al.,* 1995).

Table 16.1: List of Some Non Timber Forest Products in Arunachal Pradesh

Scientific Name	Family	Parts used	Used for
Calamus flagellum	Arecaceae	Young shoots	Used for vegetable and furniture making
Dendrocalamus hamiltonii	Poaceae	Young shoots	Edible purpose
Imperata cylindrical	Poaceae	Leaves	Thatch
Calamus erectus	Arecaceae	Leaves	Thatch
Livistonia jenkinsiana	Graminae	Leaves	Thatch
Piper longum	Piperaceae	Leaves, fruit	Vegetable/Spice
Piper mullesua	Piperaceae	Leaves, fruit	Vegetable/spice
Rubia cordifolia	Rubiaceae	Stem	Dye
Bixa orelina	Bixaceae	Fruit	Dye
Taxus baccata	Taxaceae	Leaves	Medicinal purposes
Solanum nigrum	Solanaceae	Tender shoots/ fruit	Vegetable
Mentha piperata	Lamiaceae	Whole plant	Vegetable/medicine
Coptis teeta	Ranunculaceae	Whole plant	Medicinal purpose

NTFPs supply raw materials for large scale industrial processing, including for internationally traded commodities such as foods, beverages, confectionery, flavours, perfumes, medicines, paints or polishes. Presently, 150 NTFPs are significant in terms of international trade, including honey, resin, bamboo shoots, cork, forest nuts, edible mushrooms, essential oils, plant parts, medicinal and aromatic plants. A list of important NTFPs occurring Arunachal Pradesh which have greater potential for income generation is given in the Table 16.2.

NTFPs and Income Generation

Some of the collected NTFPs are sold in the market to earn some income from the selling of these forest products. Traders and middlemen usually buy these products from the tribals and take opportunity to earn most of the benefit. But the tribal people in the remote area are getting minimum benefit instead of their hard work. There is an unemployment problem in the country particularly in the rural area. A number of NTFP have high market value and are being commonly sold in local or outside markets. So there is a great scope to make some permanent NTFP base market and establish proper marketing links to promote permanent revinew collection. To fulfill the market demand cultivation of high value NTFP species is required immediately. In this process a large number of people will get absorbed and get benefited. Plantation, tending, weeding, pruning, harvesting, processing and marketing etc., can generate sufficient job opportunities for the rural tribes. Since the most of the population of Arunachal Pradesh are intermingling with forest produce directly or indirectly they can also employed in production and collection of NTFPs as in India as worked out by Gupta and Guleria (1984) mentioned below in the Table 16.2.

Table 16.2: Current and Potential Production and Employment in Collection of NTFPs in India

Description	Collection Period	Production Potential (MT)	Employment Potential (Man-year)
Fibres and flosses			
Fibre	March-May	45000	79000
Kapok floss	May-June	4500	1500
Grasses	Oct-March	525000	1800000
Bamboo and canes			
Bamboo	All year round	4309000	110000
Canes	All year round	21000	1050
Essential oils			
Lemon grass	May-June	1950	32550
Palmarosa	October-November	135	2250
Eucalyptus	All year round	210	3480
Cinnamommum	All year round	4	70

Contd...

Table 16.2–Contd...

Description	Collection Period	Production Potential (MT)	Employment Potential (Man-year)
Sandal	All year round	225	2250
Deodar	All year round	23	230
Pine	All year round	100000	100000
Non-edible oil seeds			
Mahua	April-June (northern) Oct-Nov (southern)	490000	163000
Neem	May-June	418000	70000
Karanj	June-Oct.	111000	37000
Kusum	June-July	90000	30000
Sal	April-June	5504000	1123000
Kokum	May-June	2000	700
Nahor	May-June	46300	15300
Undi	April-June and Sep-Nov	3800	
Tans and dyes			
Babul bark	All year round	50000	8300
Avaram bark	All year round	45000	7500
Wattle bark	All year round	45000	7500
Myrobalans	January-March	150000	9900
Gums and resins			
Karaya gum	April-June	22500	75000
Ghatti and babul	April-June	3000	10500
Resins	March-June	150000	60200
Lac and tasar silk			
Lac and lac products	Oct-Jan and April-July	30000	10950
Tasar silk	Aug-Dec	1900	9500
Tendu Leaves	April-June	300000	107000
Drugs, Spices and insecticides			
Sarpagandha	Not available	1600	42670
Kuth	Oct	1000	26670
Cinchona	Not available	2000	3335
Edible products	Not available	N.A	N.A

Source: Gupta and Guleria (1984).

NTFP and Socio-Economic Development

NTFPs are the most important resource for upliftment of economy of rural people. For the rural development, the locally available resources have to be thoughtfully and systematically harnessed. Different forest related activities may also generate employment and income through many development schemes such as match splints manufacture, rope making, extraction of tanning material, manufacturing unit of bamboo and cane articles, furniture units, manufacture of tools handles and other small implements, manufacture of tooth picks, distillation of essential oils such as Eucalyptus, Citronella etc., bee keeping, lac propagation, tassar reeling, resin tapping, starch making, hand paper manufacture, making leaf plates, wood carving, charcoal making, broom making, katha manufacture, collection and processing of medicinal plants, agarbatti manufacture, repairing units of wooden/agricultural implements, fruit canning and food processing units. A number of other small scale and cottage industries may further be thought to provide self employment to the rural people.

Arunachal Pradesh has tremendous prospects for NTFPs based Industries. Some of them are:

1. Cottage industry
2. Paper and pulp industry
3. Honey and Bee wax
4. Turpentine oil and Resin
5. Orchid cultivation
6. Medicinal and aromatic plants cultivation
7. Mushroom cultivation.

Some important measures those could be taken for the upliftment of socio-economic status are:

☆ Government, NGOs and assistance agencies should place adequate emphasis on NTFPs activities, which have a high potential for poverty alleviation.

☆ Government should implement designing and evaluating development policies and programs for rural people.

☆ All concerned institutions and organizations should be dedicated to enhance the overall socio-economic benefits based on NTFPs.

☆ National government, NGOs, international agencies, research institutions and universities should give attention to the value of all benefits of NTFPs (including specific non-market benefits and socio-cultural values).

☆ All concerned agencies and institutions should help to raise the awareness of policy-makers and planners on the real significance and importance of NTFPs.

☆ International and national development agencies should ensure that the environment, economic and socio-cultural interests are adequately represented in the teams for management of NTFPs.

Market Scenario of NTFPs

For economic upliftment of the rural people and to enjoy their rights, the illegal interference of middleman and traders in the market of NTFPs must be eliminated. The most important factor is good market networking for NTFPs available at the remote area.

The markets of NTFPs are not well established and not steady. For example *Coptis teeta* was fetching a good price around Rs 600 to 750 per kg in Tezu and Roing earlier, while at the same time in Kolkata market price was Rs 1200 to 1500 per kg (Sundriyal R.C. *et al.,* 2002). International market for medicinal plants is US $ 60 billion per year that is growing at the rate of 7 per cent. Presently India export herbal material and medicines to the turn of only US $ 100 million.

Table 16.3: Market Price of Industrial Bamboo in Various States

Sl.No	Market	Price per SU/NT/MT
1.	Bangalore	1798 to 1868 per MT
2.	Calcutta	1200 per ton
3.	Hyderabad	110 to 1265 per MT
4.	Jabalpur	2250 per MT
5.	Madras	465 per MT
6.	Nagpur	1800 per MT

SU: Sale Unit; NT: Notional Ton; MT: Metric Ton.

Source: National Data Base of Bamboo (INDIA), (1995), ICFRE/INBAR/95/1

Bamboos are abundant in Arunachal Pradesh and they are marketed in two forms, such as commercial bamboo and industrial bamboo. The dry and green bamboo pieces of 1 to 2 meter length is called industrial bamboo and provide for paper mills. The market prices of industrial bamboo in some states are given in Table 16.3.

Generation of Awareness for Sustainable Harvesting of NTFPS among Rural People

The NTFPs are the minor revenue generating products for rural people. It is necessary to create awareness on conservation and sustainable uses of NTFPs. It is therefore, necessary to study the impact of harvesting NTFPs on their regeneration potential in order to advocate the optimum level of harvesting for maintaining ecological sustainability. For increasing realization for ecological, socio-cultural and economic dimensions of forests, the sustainable forest management has come to be reckoned as most important intervention for sustainable development. Over harvesting of NTFPs has negative impact on many forest ecosystems of India. A number of NTFPs are destructively harvested by pre –mature plucking of fruits, seeds, roots, rhizomes, leaves etc.Unsustainable and destructive harvesting decrease the forest resource. Sustainable NTFP management is an important for reduce the destructive harvesting by the rural tribal people.

Conclusion

Tribal populations totally depend upon NTFPs for income and subsistence. These are also significant source of subsistence products, employment and household income in rural area of tribal dominated states of India including Arunachal Pradesh in North-East. The uses of NTFPs are high in Arunachal Pradesh, because the great majority of population of this state is directly or indirectly depends upon forest products. It has been observed that the tribal people are exploited by middlemen when buying the NTFPs collected by them. Therefore, different types of tribal cooperative societies have to be formed to develop their own economy. There is an ardent need from government side to protect their benefits by providing them strong marketing network. Tribal societies themselves should come ahead to safe guard the benefits for the tribals.

References

Baishya, A.K., Balodi, B. and Pal, G.D., 2002. Floristic diversity of Arunachal Pradesh: An overview. In: *Arunachal Pradesh Environmental Planning and Sustainable Development: Opportunities and Challenges*, (Eds.) Sundriyal *et al.*

Chauhan, J.A., 2002. Prospects and sustainable utilization of non-timber forest products in Arunachal Pradesh. In: *Arunachal Pradesh Environmental Planning and Sustainable Development: Opportunities and Challenges*, (Eds.) Sundriyal *et al.*

Arun Dwivedi, A.P., 1993. *Forest: The Non-Timber Resources.* International Book Distributor, Dehradun, India.

Gupta, T. and Guleria, A., 1984. *Non-Wood Forest Products of India.* Oxford and IBH Publication, New Delhi.

Haridasan, K., Shukla, G.P. and Beniwal, B.S., 1995. In: *Medicinal Plants of Arunachal Pradesh.* State Forest Research Institute, Itanagar.

Roy, P.S. and Behera, M.D., 2005. Assessment of biological richness in different altitudinal zones in the Eastern Himalayas, Arunachal Pradesh, India. *Current Science*, 88(2): 250–257.

Theagarajan, K.S., 1994. Non-wood forest products: Problems and perspectives. *My Forest*, 30(1):33–35.

Chapter 17

Some Common Medicinal Plants Used by the Samuguria Mishing Tribe of Lakhimpur District of Assam

Annajyoti Gogoi, Jully Gohain (Neog), Tulika Gogoi and Prabhat Saikia

ABSTRACT

The Samuguria Mishing tribe forms a major indigenous tribe of Lakhimpur district. They are a distinct section of the Dahgam section.They have a unique position in the larger Mishing tribe as they speak only Assamese language and not the Mishing dialect. They mostly settle near the bank of rivers. They use plants for their traditional medicinal purposes. Information regarding the uses of the plants were consulted with the local healers, village elders, housewives and the local men.The present study aimed to prepare an inventory of the medicinal plants used by them in various ailments like blood pressure, diabetes, tuberculosis, sexual disorder, stomach troubles, fever,etc.The common plants belong to families like

Rubiaceae, Verbenaceae, Solanaceae, Urticaceae, Asteraceae, Orchidaceae, Euphorbiaceae, etc.Vernacular names along with families and uses are discussed in full length of the paper.

Keywords: *Samuguria Mishing, Traditional, Inventory, Medicinal plant.*

Introduction

The Lakhimpur district lies between 26°48' and 27°53' N latitude and 93°42' and 94°20' W longitude. The Mishing tribe is an important tribe of the district. Of the eleven clans found in the tribe, Samuguria is a major clan. They constitute a distinct section of Dahgam section. They have all similarities with the Mishing tribe but exceptional in the case that they do not speak Mishing dialect.However, a few Mishing words are used by them in their daily conversation. They do not celebrate the Ali- Aye- Ligang, the main festival of the larger Mishing community and celebrate Bihu like the Assamese people. However their home garden pattern, housing pattern is same like Mishing people. They live in Chang ghar, a thatched house raised on bamboo stilts. Chai Apoong (rice bear) is the most essential in their food style for serving guests and in other religious occasions.Rice in torapat known as *Purang apin*, dry and powdered fish called *Namshing*, dried pork and chicken are their important dishes. Dry rice called *sera dia chawal* and various types of rice cakes are also common in their food items. Their primary occupation is agriculture. Although the Samugurias follows the Mishing religion, they are Hindus and they belong to the *Vaishnava* religion of Srimanta Sankardeva.In due course of time, other religions like *Krishnaguru, Ek Sarania Bhagavati Dharma, Anukul Thakur* and *Sai Baba* are also prevailing.

Methodology

The study was conducted during July 2009- 2010 in two villages Deobill and Bamundoloni which are about 9 and 15 km away from North Lakhimpur. Both the villages are situated on the bank of the river Ranganadi. To get information regarding the uses of different plants, personal interviews with the village people like local healers or 'kabiraj', gaon buras, elderly persons, housewives, etc were made. The local markets were visited at frequent intervals to collect information regarding their uses. Collected plants were identified in the field itself or consulted with Flora of Assam, Vol 1- 4 (Kanjilal *et al.*, 1934 - 41); Vol 5. (Bor, 1940) and Assam's Flora (Choudhury, 2005).During the survey, local name of the plants as far as possible along their uses were recorded.

Results

A total of 60 plants species belonging to 43 families were recorded to be used by the tribe in different ailments. Among the families, Rubiaceae is the most dominant followed by Euphorbiaceae, Poaceae, Mimosaceae, Liliaceae, Acanthaceae, Apiaceae, and so on (Table 17.1). The plant species were arranged alphabetically with their botanical names, vernacular names, families and medicinal uses (Table 17.2).

Table 17.1: List of Families in Increasing Order of Occurrence

Sl.No.	Family	Per cent of Occurence
1.	Rubiaceae	4
2.	Euphorbiaceae	3
3.	Myrtaceae	3
4.	Poaceae	3
5.	Acanthaceae	2
6.	Apiaceae	2
7.	Combretaceae	2
8.	Liliaceae	2
9.	Menispermaceae	2
10.	Moraceae	2
11.	Orchidaceae	2
12.	Verbenaceae	2
13.	Anacardiaceae	1
14.	Apocynaceae	1
15.	Araceae	1
16.	Aristolochiaceae	1
17.	Asteraceae	1
18.	Balsamiaceae	1
19.	Bromeliaceae	1

Contd...

Table 17.1–Contd...

Sl.No.	Family	Per cent of Occurence
20.	Caesalpiniaceae	1
21.	Caricaceae	1
22.	Caryophyllaceae	1
23.	Clusiaceae	1
24.	Commelinaceae	1
25.	Cuscutaceae	1
26.	Cyperaceae	1
27.	Lauraceae	1
28.	Lythraceae	1
29.	Meliaceae	1
30.	Mimosaceae	1
31.	Musaceae	1
32.	Oleaceae	1
33.	Oxalidaceae	1
34.	Papaveraceae	1
35.	Papilionaceae	1
36.	Piperaceae	1
37.	Plumbaginaceae	1
38.	Polygonaceae	1
39.	Rosaceae	1
40.	Rutaceae	1
41.	Zingiberaceae	1
42.	Solanaceae	1
43.	Urticaceae	1
	Total	**60**

Discussion

Today about 4.4 billion people comprising 80 per cent of world's population rely on plants as their primary source of medicine. Our country India is also inhabited by over 550 tribal communities belonging to 227 ethnic groups constituting 7.7 per cent of the entire population. About 10,000 plants are traditionally used all over the

Table 17.2: List of Medicinal Plants Used by the Samuguria Mishing Tribe of Lakhimpur District

Sl.No.	Name of the Species	Vernacular Name	Family	Uses	Parts Used
1.	*Abrus precatorius* L.	Latumoni	Papilionaceae	Tonsillitis	Fruit
2.	*Acorus calamus* L.	Boch	Araceae	Dysuria	Rhizome
3.	*Allium sativum* L.	Naharu	Liliaceae	Dryusuria,cough with fever, infantile fever, anaemia	Bulb
4.	*Alstonia scholaris* R.Br.	Chatian	Apocynaceae	Anaemia, astringent	Bark, Leaves
5.	*Andographis paniculata* (Burm.f) Wall	Kalpatita	Acanthaceae	Fever, cough, gastric ulcer, malaria	Leaves
6.	*Ananus comosus* (L.) Merr	Annaros	Bromeliaceae	Diarrhoea, blood dysentery	Fruit
7.	*Aristolochia indica* Linn	Isharmul	Aristolochiaceae	Female sterility, blood dysentery, dry cough	Bark, Leaves, Fruit
8.	*Artocarpus chama* Buch Ham	Chama kathal	Moraceae	Morbid	Bark, Leaves and fruit
9.	*Asparagus racemosus* Wild.	Satmul	Liliaceae	Leucorrhoea, anaemia, Stomach ache	Root, Stem
10.	*Azadirachta indica* A. Juss	Mahanim	Meliaceae	Loss of appetite, stomach ache, chest pain, diabetes	Leaves
11.	*Bambusa tulda* Roxb.	Jati baah	Poaceae	Haemostatic	Young shoot
12.	*Caesalpinia bonduc* (L.) Roxb.	Latagutti	Caesalpiniaceae	Gastric complaints, fever	Fruit

Contd...

Table 17.2–Contd...

Sl.No.	Name of the Species	Vernacular Name	Family	Uses	Parts Used
13.	*Carica papaya* L.	Omita	Caricaceae	Antifertility, indigestion, against tapeworms	Pulp of unripe fruit, flowers of male plant
14.	*Centella asiatica* L.	Manimuni	Apiaceae	Gripping pain after child birth, gastric complaints, tuberculosis, cholera, anaemia	Leaves, root
15.	*Cissampelos pareira* L.	Tubukilota	Menispermaceae	Sexual disorder, cuts and wounds, indigestion and relieving urinary irritation	Leaves
16.	*Clerodendron colibrookianum* Walp.	Nafafu	Verbenaceae	Hypertension, morbid	Leaves
17.	*Clerodendron indicum* (L.) O. Kuntge	Dhapattita	Verbenaceae	Stomach ache, anthelmintic cough	Leaves
18.	*Coix lachnyma- jobi*	Kaurimoni	Poaceae	Waist pain	Root
19.	*Commelina benghalensis* L.	Kona himolu	Commelinaceae	Tuberculosis, Infection of eyes	Juice of young stem
20.	*Costus specious* (Koen.) Smith	Jomlakhuti	Zingiberaceae	Gripping stomach pain of women, dyrusia and pneumonia	Juice of root
21.	*Croton joufra* Roxb.	Mahudi	Euphorbiaceae	Pneumonia	Juice of leaves
22.	*Cuscuta reflexa* Roxb.	Akashilota	Cuscutaceae	Anaemia	Stem
23.	*Cynodon dactylon* (L.) Pers.	Duboribon	Poaceae	Dyrusia	Juice of leaves
24.	*Cyperus rotundus* L.	Keabon	Cyperaceae	Cough	Juice of leaves

Contd...

Table 17.2–Contd...

Sl.No.	Name of the Species	Vernacular Name	Family	Uses	Parts Used
25.	*Dentella repens* Frost.	Matipat	Rubiaceae	Lochial discharge, stomach ache	Leaves
26.	*Drymaria cordata* Wild	Laijabori	Caryophyllaceae	Opacity	Juice of leaves
27.	*Eclipta alba* Haesk.	Kenharaj	Asteraceae	Haematuria, stomach ache during menses	Leaves
28.	*Entada pursaetha* DC.	Gila	Mimosaceae	Cirrhosis	Fruit
29.	*Euphorbia hirta* L.	Sakhiroti bon	Euphorbiaceae	Menstrual complaints	Leaves
30.	*Garcinia cowa* Roxb. ex DC.	Kujithekera	Clusiaceae	Cholera	Fruit
31.	*Hydrocotyl rotundifolia* Roxb.	Bormanimuni	Apiaceae	Cholera, anaemia, blood dysentery, Stomach pain during menses	Leaves
32.	*Impatiens balsamia* Linn.	Koriabjol, Dum-deuka	Balsamiaceae	Tonsillitis, Fever, Fracture	Leaves, stem
33.	*Lawsonia inermis* L.	Jetuka	Lythraceae	On wounds as antiseptic, in infections of nails of toes	Leaves
34.	*Litsea monopetala* (Roxb.)	Hoanlu	Lauraceae	Gastric complaints, rodent ulcer	Root
35.	*Musa balbisiana* Colla.	Vimkol	Musaceae	Blood dysentery, as antiseptic	Fruit
36.	*Mussaenda roxburghii* Hook.	Sukloti	Rubiaceae	Infantile diarrhoea, dysentery	Leaves
37.	*Nyctanthus arbor- tristis* L.	Sewaliphool	Oleaceae	Malaria, fever	Flower, Leaves

Contd...

Table 17.2–Contd...

Sl.No.	Name of the Species	Vernacular Name	Family	Uses	Parts Used
38.	*Oldenlandia corymbosa* Linn.	Bonjaluk	Rubiaceae	Dysentery, fever, pain, gastric, jaundice, nerve problems	Whole plant
39.	*Oxalis corniculata* L.	Tengese tenga	Oxalidaceae	Cholera	Whole plant
40.	*Paedaria foetida* L.	Bhedailota	Rubiaceae	Tuberculosis, menstrual complaints, stomach pain	Leaves, stem
42.	*Papilionthae terres* (Roxb.) Schltr	Bhatowphool	Orchidaceae	Fracture, neuralgia	Stem
43.	*Phlogocanthus thyrsiformis* (Hardw.) Mabberley	Titaphool	Acanthaceae	Fever, stomach ache, cough, dysentery, bronchitis	Flowers, leaves
44.	*Piper longum* L.	Pipoli	Piperaceae	Antirabies in dog bite	Leaves
45.	*Plumbago zeylanica* L.	Boga agechi	Plumbaginaceae	Cirrhosis	Leaves
46.	*Polygonum hydropiper* L.	Beholongni	Polygonaceae	Gastric complaints, chest pain, Liver problems, cholera	Leaves, young shoot
47.	*Psidium guajava* L.	Madhuriam	Myrtaceae	Blood dysentery, early removal of placenta, Tonsillitis	Fruit, Leaves
48.	*Rhynchostylis retusa* (L.)Bl.	Kopouphool	Orchidaceae	As emollient and supplied as drops in otorrhoea	Juice of leaves
49.	*Riccinus communis* L.	Ara	Euphorbiaceae	Snake bite, Post natal pain	Leaves, Root
50.	*Rubus hexagynous* Roxb.	Jutulipoka	Rosaceae	Piles	Juice of root

Contd...

Table 17.2–Contd...

Sl.No.	Name of the Species	Vernacular Name	Family	Uses	Parts Used
51.	*Sarcochlamys pulcherrima* (Roxb.) Gourd	Mesaki, Ombe	Urticaceae	Fever, diabetes	Young Leaves
52.	*Solanum torvum* Sw.Prodr.	Hatibhekuri	Solanaceae	In spleen enlargement	Fruit
53.	*Spondias pinnata* (L.f.)Kurz	Amora	Anacardiaceae	Cholera	Fruit, Leaves
54.	*Streblus asper* Lour	Kharua	Moraceae	Snake- bite, fever, sinusitis	Root, Latex
55.	*Syzygium aromaticum* (L.) Merr.	Long	Myrtaceae	Gastric complaints, tooth ache, chest pain, liver problems & cholera	Fruit and oil
56.	*Syzygium cumini* (L.)Skeels	Kolajamu	Myrtaceae	Diarrhoea, stomach ache	Fruit, Bark and leaves
57.	*Terminalia bellilrica* (Gaertn) Roxb.ex.Flem.	Bhumura	Combretaceae	Blood purification, gastric ulcer	Fruit
58.	*Terminalia chebula* Retz	Hilikha	Combretaceae	Gastric ulcer, stomach ache, Diabetes	Fruit
59.	*Tinospora cordifolia*(Willd.) Hook.f.&Th.	Hidhilota	Menispermaceae	Chronic dysentery, waist pain, cuts and wounds, fractures, malaria and gonorrhia	Juice of the plant
60.	*Zanthoxylum nitidum* (Roxb.)	Tejmuri	Rutaceae	in menstrual complaints, toothache, gastric complaints	Juice of young shoot, stem and root

country in various purposes (Goel, 2009).In our present study also about 60 plant species are found to be used by the Samugurias for various types of ailments and these plants have indicated their potential medicinal properties. A similar study was also done by Yonggam(2005) on the Mishing tribes of Arunachal Pradesh.Some of the common ailments that were cured are fever, stomach pain, anaemia, drysuria, cough, diarrhoea, gastric complaints, cuts and wounds, gastric complaints, fractures, anaemia, snake bites, tonsillitis,malaria,etc. However, mention should be made of the plant species like *Allium sativum, Centella asiatica, Commelina benghalensis* and *Paedaria foetida* which are used against tuberculosis.In dysentery various plants like *Ananus commusus, Aristolochia indica, Hydrocotyl rotundifolia, etc* are used. A variety of plants are also used in sexual disorders like female sterility, gripping pain after child birth, early removal of placenta,stomach ache during menses and other menstrual complaints. *Rubus hexagynous* was used against piles. *Azadirachta indica* was used against diabetes.*Piper longum* was used as antirabies in dog bite.

Conclusion

Tribal societies of various parts of the world use enormously various wild plants for medicinal purposes. Our country is considered as a heritage of traditional system of medicine dating back to 5000 BC (Goel, 2009).From our study we can conclude that the Samugurias also have a rich stock of information regarding the herbal medicine. They have used these traditional medicinal plants antiquity and is passed on verbal tradition to the younger generation. However, due to impact of urbanization and semi modernization, this system of usage of plants is gradually fading from the young generation. There is an urgent need for conservation and domestication of these valuable plants for further studies.

References

Bor, N.L., 1934–1940. *Flora of Assam*, Vol. 5.

Choudhury, S., 2005. *Assam's Flora*. A.S.T.E.C., U.N.Bezbaruah Road, Silpukhuri,Guwahati, Assam.

Goel, A.K., 2009. Conservation, value addition and commerciali- zation of traditional knowledge in India paper presented in *National Conference on Recent Trends in Biodiversity Researches,*

Organized by Department of Life Science, Assam University, Silchar in collaboration with Society of Ethnobotanists, Lucknow and Association of Plant Taxonomy, Dehradun on 16–18 March.

Jain, S.K., 1987. *Manual of Ethnobotany,* Scientific Publications, Jodhpur.

Kanjilal, V.N., Kanjilal, P.C., Das, A., De, R.N. and Bor, N.L., 1934–1940. *Flora of Assam in 5 Vols.* Govt. Press, Shillong.

Mipun, J., 1974. *The Mishings of Assam.* Gian Pub. House, New Delhi.

Rawat, M.S. and Shankar, R., 2005. Folk medicinal plants in Arunachal Pradesh.*Arunachal Forest News,* 21(1 and 2): 59–64.

Saha, D. and Dutta, B.K., 2001. Some observations on the status of medicinal plants of Barak Valley, Assam paper presented in *National Seminar on Ethno-medicine of North East India* published by National Institute of Science Communication and Information Resources, CSIR, New Delhi, India.

Yonggam, D., 2005. Ethno-medico botany of the Mishing tribe of East Siang District of Arunachal Pradesh. *Arunachal Forest News,* 21(1 and 2): 44–49.

Chapter 18

Traditional Treatment of Diarrhea and Dysentery with Phytomedicine Among Tea Tribes: A Case Study of Tinsukia District of Assam

Deepjyoti Saikia and Chandan Das

ABSTRACT

Plants have been used for medicinal purposes to treat and prevent illness for as long as history has been recorded. China, India, Egypt and Assyria appear to have been the places which cradled the use of herbs. Despite the progress in orthodox medicine, the practice of phytomedicine is now flourishing today as the primary form of medicine in the world. The indigenous people of Assam including the Tea Tribe community have splendid tradition of healthcare system based on plants and plant products used in the preparation of phytomedicine. The Tea Tribes is now one of the ethnic groups of Assam was brought in by colonial planters as indentured labourers from Chotanagpur, Madhya Pradesh and Orissa.

This paper documents the traditional knowledge of medicinal plants, processes that are used for the treatment of diarrhea and dysentery by the Tea Tribes of three different Tea Estate viz. Chandmari, Kachujan, and Sukanpukhuri (Sikajan) of Tinsukia district of Assam. For this study the methods employed were field survey, direct interactions with the village headman, other villagers of different age group and sex, kabiraj and medical officers. The ethno-medicinal plants used for their treatment of diarrhea and dysentery along with their botanical name, family, plant parts and processes are reported.

Keyword: *Orthodox medicine, Phytomedicine, Tea tribe, Ethno-medicinal plants, Diarrhea and dysentery.*

Introduction

The history of plant medicine dates back perhaps from the origin of human. In India the use of plant medicine known from the *Rigveda* in very brief the period is estimated between 3500 – 1800BC. A little more detail available in the *Atharvaveda*. After that two most important work of Indian traditional medicine, the work of *Charak* and *Susruta*, the *Charak-Samhita* and *Susruta-Samhita* written around 100AD and 2500BC. Since then the use of plants in curing and healing is as old as man himself and still civilization depends upon nature. Nowadays the ethnobotany is a interdisciplinary subject like pharmaceutical science, medicine, etc, so it is now known as ethno-medicinal science.

Phytomedicine is the use of plants, parts of plants and isolated phytochemicals for the prevention and treatment of various health concerns. Phytomedicine or herbalism is usually defined as an alternative medicine, the therapeutic use of extracts from flowers, fruits, roots, seeds, barks, stems, leaves, alone or as an adjunct to other forms of alternative health care or physical manipulation *e.g.* massages.

Diarrhea is defined by the W.H.O. as having three or more loose or liquid stools per day or as having more stools than is normal for that person. Gastroenterologists define as the passage of more than 200 gm of stool daily (for an adult or typical western diet). Diarrhea can be further divided as –Acute diarrhea, Persistent diarrhea and chronic diarrhea according to period of infection. The causes of diarrhea mainly due to Bacteria like *Bacillus, Staphylococcus, Clostridium, Schizella, etc.,* Virus such as *Rotavirus, Norovirus,* and

Protozoan like *Giardia, Cryptosporidium, Microsporidium, Isosporium,* etc. The Dysentery is an inflammatory disorder of intestine especially of colon that results in severe diarrhea containing mucus and or blood in faeces with fever and abdominal pain. Dysentery may be Bacillary dysentery which is caused by *Shigella* group bacteria and Amoebic dysentery caused by *Entamoeba histolytica* a protozoan.

It is a common cause of death in developing countries and the second most common cause of Infant death world wide. In 2009 diarrhea and dysentery was estimated to have caused 1.1 million death in people aged below 5 years, 1.5 million death in children above 5 years old.

Study Area

Tinsukia is occupied with a unique position among the towns of Assam. Its main importance is industrial, commercial development and good transportation systems particularly roadways and railways. Tinsukia is situated between 27°23'–27°50' latitude and 95°22–95°40' longitude. It is a flat terrain without any hill with some low lying and marshy areas. The river Tingrai is passing almost along the southern boundary and in the northern side Dibru River. The soil is clay, light yellowish and dark grayish in color. The climatic condition is mesothermal muggy climate with maximum temperature up to 39° C and minimum of 8.8° C. The average annual rainfall is 28.48 c.m. As per 2001 census the total population of Tinsukia was 1150062 and the literacy rate percentage was 60.95 per cent.

The present study area is located northern part of Tinsukia which is mainly comprises of tea gardens, grassland, forest, agricultural and residential land.

Tinsukia was originally the capital of Matak Kingdom was known as Bengmara. Sarbananda Singha was the last king of Matak Kingdom. The name Tinsukia was derived from the "Tinkunia Pukhuri" which was dug by the Barborua of king Sarbananda Singha.

The so-called Tea Tribes of Assam were brought in by colonial planters (British) as indentured labourers from Chota Nagpur, Madhya Pradesh, Orissa, Jharkhand etc..In Assam they are mainly found in the districts of Darang, Sonitpur, Nagoan, Jorhat, Golaghat, Dibrugarh, Cachar, Hailakandi, Tinsukia, and almost all the districts of Assam in India. Tea tribes are one of the backward and most exploited tribes in Assam, though newer generations are comparatively educated and now have intellectuals and professionals in various fields. The Tea tribes, being basically labours, live in villages, inside tea estates located in interior places. Poor education, poverty, addiction of males to country beer, poor standard of living and health facility are the problems in their life.

The ethnobotanical knowledge of Tea tribes is very rich. In their daily life they used a large number of plants as a food and medicine, preparation of liquor, house building materials and other household activities. They always prefer the phytomedicine because of their economic condition and distance is away from the town. The knowledge of phytomedicine which they practice came from forefather as a trial method without any documentation or proper written procedure. The knowledge regarding use of herbal medicine is usually restricted to some people who are popularly known as bej and ojha.

There are certain factors for which the tea tribes prefer phytomedicine are as follows:

1. Phytomedicine has no or little side effect.
2. Low cost and easily available of phytomedicine.
3. Strong faith on local herbal medicine.

Methodology

The study was carried out during 2009–2010 in the Tea garden area and tea tribe inhabited villages were visited to meet the men and women who used phytomedicine. The information have been collected from the village chief (Gaon burahs), medicine man, and even local man and women, and cultivators using semi structured questionnaires. Information about the plants were recorded with regards to their vernacular names, plant part used, process of preparation of medicine either individually or in combination with other plant parts, and mode of application and doses for the treatment of diarrhea and dysentery. The specimens were collected and herbarium specimens were deposited in the Department of Botany, Duliajan College, Duliajan. Collected plant species were identified with the help of "Flora of Assam" (Kanjilal *et al.*, 1934–1940) and "Weeds of North East India" (Islam 1996).

Table 18.1: List of Ethno-medicinal Plants Recorded during the Study Period

Sl.No.	Botanical Name with Family	Local Name	Plant Part Used	Method Used
1.	*Acorus calamus* L. Aracaceae	Bach. (Mitphall)	Rhizome	2 – 3 teaspoonful juice of rhizome with black salt take orally twice a day until cure for both dysentery and diarrhea
2.	*Aegle marmalos* L. Rutaceae	Bel.	Bark, fruit and leaf	Sufficient amount of ripe or half ripe fruit juice or boiled unripe fruit juice used as drink with small amount of sugar and pinch of salt for few days in both dysentery and diarrhea.
3.	*Artocarpus heterophyllus* L. Urticaceae	Kothal (Kathal)	Root	Fresh root extract 2–3 spoonfuls take twice a day until cure.
4.	*Alstonia scholaris* R.Br. Apocynaceae	Chatiyana (Chaitan)	Bark	The bark juice taken in empty stomach for few days in both dysentery and diarrhea.
5.	*Albizzia labeck* Benth. Fabaceae	Sirish	Bark and Seed	The bark and seed extract used in piles and diarrhea.
6.	*Amaranthus tricolor* L. Amaranthaceae	Marisha (Lalbhaji sak)	Whole plant	The whole plant boil and used as vegetable for dysentery and diarrhea.
7.	*Amorphophallus campanulatus* Blume and Decane. Araceae.	Olkachu	Rhizome	The rhizomatous stem is given to take as vegetable with rice daily for a week.
8.	*Annona reticulata* L. Annonaceae	Atlas.	Ripe fruit	The fruit juice take daily until cure.
9.	*Asparagus racemosus* Willd. Liliaceae.	Shatamul (Bhutson)	Root	Root extract used twice a day until cure.
10.	*Bouhinia purpurea* L. Fabaceae	Rakta kanchan (Konear)	Bark	The fresh sufficient amount of bark juice is used as a drink for few days.
11.	*B. racemosa* Lamk. Fabaceae	Boga kanchan (konear)	Bark	— Do —

Contd...

Table 18.1–Contd...

Sl.No.	Botanical Name with Family	Local Name	Plant Part Used	Method Used
12.	*Bombax malabarium* Bombacaceae	Simaloo (Simal)	Root and Bark	Root and bark juice is used in dysentery and diarrhea.
13.	*Butea monosperma* Fabaceae	Palas	Gum	Two to three spoonful of gum extract is used twice a day until cure.
14.	*Calotropis gigantia* R.Fr. Ascleapiadaceae	Akon	Root and Bark	Root and bark juice is used in dysentery.
15.	*Centella asiatica* L. Apiaceae	Bormanimuni (Bengpata)	Whole plant	Entire vegetative plant crushed and 3–4 teaspoonful juice is given to take twice daily until cure.
16.	*Cinnamomum tamala* Nees and Ebern. Lauraceae	Tezpata	Leaf	Young leaves extract used twice a day until cure.
17.	*Citrus limettioides* Tanaka.	Gulnemu Rutaceae	Fruit	Preserve ripe fruit in salt and 2–3 teaspoonful of juice take twice daily or it may be take sufficient amount of fresh fruit juice with black salt until cure.
18.	*Citrus lemon* Burn. Rutaceae	Bornemu	Fruit	— Do — or Roasted fruit take twice a day until cure indysentery and diarrhea.
19.	*Citrus medica* var. *medica proper* L. Rutaceae	Jora tenga	Fruit	— Do —
20.	*Corchorus capsularis* L. Tiliaceae	Titapat	Leaf	The leaf infusion take twice daily until cure.

Contd...

Table 18.1–Contd...

Sl.No.	Botanical Name with Family	Local Name	Plant Part Used	Method Used
21.	*Cynodon dactylon* Pers. Poaceae	Dubari bon	Whole plant	The whole plant crushed and 2 spoonful extract mixed with 1 spoonful of lime juice take daily in the morning time until cure.
22.	*Datura metel* L. Solanaceae	Dhatura	Leaf and root	The leaves and root extract used in diarrhea.
23.	*Emblica officinalis* Gaertn. Euphorbiaceae	Amla or amlakhi	Fruit	Dried fruits with salt take 4–5 times daily until cure. Fresh fruit juice with citrus juice takes in sufficient amount in dysentery for few days.
24.	*Euphorbia hirta* Linn. Euphorbiaceae	Gakhirati bon (Dudh bon)	Whole plant	The whole plant crushed and take the extract of 2–3 spoonfuls for two times daily in dysentery.
25.	*Garcinia kydia* Roxb. Clussiaceae	Kuji thekera	Fruit	Sun dried slices of mature fruits infusion mixed with a little salt and 2–3 drops of mustard oil and take in sufficient amount 2–3 times daily in blood dysentery and also this infusion added the flesh of *Musa balbisiana* ripe fruit and take 2–3 times daily in dysentery until cure.
26.	*Garcinia pedunculota* Roxb. Clussiaceae	Bor thekera	Fruit	Same as in *Garcinia kydia.*
27.	*Houttuynia cordata* L. Saururaceae	Mosandari (Bisar pat)	Whole plant	The whole plant rousted with garlic and a little salt covering with a banana leaf and take with rice for few days.
28.	*Hydrocotyle sibthorpiodes* Lam. Apiaceae	Haru manimuni (Chato bengpata)	Whole plant	Entire plant crushed and 3–4 spoonful of juice with a little salt take 2–3 times daily until cure.

Contd...

Table 18.1—Contd...

Sl.No.	Botanical Name with Family	Local Name	Plant Part Used	Method Used
29.	*Melastoma malabathricum* Linn. Melastomaceae	Phutuka (Phutkon or missi)	Leaf, Flower	Leaves and flower tips are crushed and take 2–3 spoonful of juice 2–3 times daily for few days.
30.	*Musa balbisiana* Colla. Musaceae	Athiyakal (Gutikal)	Fruit and stem sap	Ripe fruits infusion with *G. kydia* in sufficient amount for 2–3 times daily until cure. Use bamboo cylinder to collect stem sap and take orally until cure.
31.	*Musa paradisiacal* L. Musaceae	Purakal (Sabjikal)	Green fruit	The boiled green fruits crushed and used as vegetable for few days.
32.	*Murraya koenigii* Spreng. Rutaceae	Nara-singha	Leaves	The boiled leaves extract used with rice for few days.
33.	*Mimusops elengi* L. Roxb. Sapotaceae	Bakul	Pulp of ripe fruit	Pulp of ripe fruit extracts in sufficient amount take 2–3 times daily for a week.
34.	*Mesua ferrea* L. Guttiferae	Nahar	Flower bud	Extract of flower buds with little black salt take 3–4 spoonful daily 2–3 times until cure.
35.	*Oroxylum indicum* Vent. Bignoniaceae	Bhatghilla (Hohen gach)	Root and bark	Fresh root and bark extract a few amount take 1–2 times for 3–4 days or until cure.
36.	*Phyllanthus niruri* Linn. Euphorbiaceae	Bon amlakhi	Whole plant	The plant boiled and crushed and the juice is taken with a little salt both orally or with rice for few days.
37.	*Paederia foetida* L. Rutaceae	Bhedai lata	Leaf	— Do —

Contd...

Table 18.1–Contd...

Sl.No.	Botanical Name with Family	Local Name	Plant Part Used	Method Used
38.	*Psidium guajava* L. Myrtaceae	Modhuri (Temross)	Young shoot, root, bark, fruit	3–4 young shoot crushed and 2–3 spoonful of extract used 2–3 times daily until cure. Root and bark extract also used similar manner.
39.	*Syzygium cumini* Skeels. Myrtaceae	Jamuk (Jamun)	Bark, leaf, fruit	Bark and leaves extract take 2–3 times daily until cure. The fruit juice in sufficient amount takes two times daily until cure.

Result and Discussion

The use of phytomedicine among the tea tribe communities in different parts of Assam is an age old practice. Overall, 39 collected plants species belonging to 21 families and 33 genera has been recorded. The collected plants are wild as well as cultivated. The three dominant families that yield phytomedicine to treat dysentery and diarrhea are Rutaceae, Fabaceae and Euphorbiaceae. From the field experiment it was observed that plant species such as *Musa balbisiana, Centella assiatica, Hydrocotyle sibthorpioides, Citrus limettioides, Garcinia kydia, Huyttrania cordata*, are used more frequently. Out of these plants the ethnobotanically important plants which are recorded during this project such as *Garcinia kydia, Oroxylum indicum, Citrus medica, Butea monosperma, Amorphophallus campanulatus* are very few in number and rare. Therefore, effective measures should be taken to conserve these plants from extinction. The main cause of extinction of these plants and other species of the study area are cutting and uprooting of plants (deforestation) for tea garden extension, establishment of small tea garden, housing, and for cultivation. The climatic condition in these area has changed (40° in 2009, Gillapukhuri; KVK) due to deforestation of the neighbouring area and reserved forest responsible for loss of plants.

Conclusion

From the above study we may conclude that tea tribes are very rich in ethno-botanical medicine and therefore actions should be accorded to:

1. Create awareness among local people about medicinal importance of particular plant species and their conservation and preservations.

2. Create awareness against deforestation and value of forest and future prospective.

3. Organize NGOs to educate them and plant valuable medicinal plants before extinction.

4. Reduce the deforestation process for establishment of tea gardens.

The ethno-medical claims reported here is merely an indication of our present state of knowledge on the traditional usage of herbal drugs by the groups studied. The findings may be relevant to modern phytotherapy and medical practice as well as trans-cultural health policies. Therefore, it is greatly needed to assess these plants for phytochemical analyses and ethnopharmacological screenings so as to validate the efficacy of indigenous herbal medicine used by the tea tribes of Tinsukia district.

References

Ahuja, B.S., 1993. *Medicinal Plants of Saharanpur: Survey of Medicinal Plants.* Central Council of Ayurvedic Research, Haridwar.

Islam, M., 1996. *Weeds of North East India, Dhaiali, Sibsagar, Assam, India.*

Mitra, J.N., 1988. *Introduction to Systematic Botany and Ecology.* The World Press Private Limited, Kitab Mahal, Kolkata.

Jain, S.K., 2004. *Medicinal Plants.* National Book Trust India, Green Park, New Delhi.

Kanjilal, V.N., Kanjilal, P.C., Das, A., De, R.N. and Bor, N.L., 1934–1940. *Flora of Assam in 5 Vols.* Govt. Press, Shillong.

Trivedi, P.C., 2006. *Medicinal Plants: Ethnobotanical Approach.* Agrobios, Jodhpur, India.

Index